Endurance Time Excitation Functions

Seismic assessment and earthquake-resistant design are essential applications of earthquake engineering for achieving seismic safety for buildings, bridges, infrastructure, and many other components of the built environment. The Endurance Time Method (ETM) is used for seismic analysis of simple and complex structural systems and civil engineering infrastructure as well as producing optimal and cost-effective structural and detail designs. ETM is a relatively new approach to seismic assessment and design of structures. It has developed into a versatile tool in the field, and its practical applications are expected to increase greatly in the near future.

Taylor and Francis Series in Resilience and Sustainability in Civil, Mechanical, Aerospace and Manufacturing Engineering Systems

Series Editor
Mohammad Noori
Cal Poly San Luis Obispo

PUBLISHED TITLES

Sustainable Development for the Americas
Science, Health, and Engineering Policy and Diplomacy
Honorary Editor: E. William Colglazier
Hassan A. Vafai and Kevin E. Lansey
With the administrative assistance of Molli D. Bryson

Endurance Time Excitation Functions
Intensifying Dynamic Loads for Seismic Analysis and Design
Homayoon E. Estekanchi
Mohammadreza Mashayekhi
Hassan A. Vafai

SERIES EDITOR

Mohammad Noori is a Professor of Mechanical Engineering at California Polytechnic State University, San Luis Obispo. He received his B.S. (1977), M.S. (1980), and Ph.D. (1984) from the University of Illinois, Oklahoma State University, and the University of Virginia, respectively, all degrees in Civil Engineering with a focus on Applied Mechanics. His research interests are in stochastic mechanics, nonlinear random vibrations, earthquake engineering and structural health monitoring, AI-based techniques for damage detection, stochastic mechanics, and seismic isolation. He serves as the executive editor, associate editor, the technical editor, or a member of editorial boards of eight international journals. He has published over 250 refereed papers, has been an invited guest editor of over 20 technical books, has authored/co-authored six books, and has presented over 100 keynote and invited presentations. He is a Fellow of ASME and has received the Japan Society for Promotion of Science Fellowship.

For more information about this series, please visit: https://www.routledge. com/Resilience-and-Sustainability-in-Civil-Mechanical-Aerospace-and-Manufacturing/book-series/ENG

Endurance Time Excitation Functions

Intensifying Dynamic Loads for Seismic Analysis and Design

Homayoon E. Estekanchi, Mohammadreza Mashayekhi, and Hassan A. Vafai

CRC Press
Taylor & Francis Group
Boca Raton London New York

CRC Press is an imprint of the
Taylor & Francis Group, an **informa** business

Designed cover image: Shutterstock

First edition published 2023
by CRC Press
6000 Broken Sound Parkway NW, Suite 300, Boca Raton, FL 33487-2742

and by CRC Press
4 Park Square, Milton Park, Abingdon, Oxon, OX14 4RN

CRC Press is an imprint of Taylor & Francis Group, LLC

ISBN: 9781032107127 (hbk)
ISBN: 9781032107141 (pbk)
ISBN: 9781003216681 (ebk)

DOI: 10.1201/9781003216681

Typeset in Times New Roman
by codeMantra

Contents

List of Figures

List of Tables

Foreword

I am delighted to write the Foreword for this book, which is the compilation of the recent advances on fundamentals of generating the "Endurance Time Excitation Functions" and their application to earthquake-resisting design of structures. The protection of buildings against earthquakes is of critical importance in seismic regions. In the conventional aseismic design, the structure is strengthened to sustain the expected earthquake ground shaking without failure and/or significant damages. However, observations of structural failures during earthquakes revealed that most destruction occurs due to the gradual increase of damages and weakening of the building due to the earthquake-induced loads. In this book, innovative methods for producing intensifying dynamic loads for performance analysis of buildings to earthquake strong-motion are presented. In this approach, the structure is subjected to a seismic excitation whose intensity increases with time. The building damages are monitored until the collapse of the structure and the corresponding Endurance Time are evaluated. The Endurance Time Method provides a new realistic approach for the aseismic design of complex structural systems.

The authors of this book are the pioneers in this area and have developed the Endurance Time methodology from its inception to its applications to the earthquake-resisting design of structures. I believe this book provides the readers with an in-depth understanding of the Endurance Time loading functions for the aseismic design of buildings. The authors introduce the fundamental aspects of the Endurance Time Excitation Functions and bring it all the way to the actual design of buildings. They also described the energy- and nonlinear-based approaches for the generation of Endurance Time Excitation Function. A more advanced generation of endurance time excitation for probabilistic seismic is presented. Optimization approaches for the application of Endurance Time methodology are also described.

The book provides an easy-to-understand description of the fundamentals of generating Endurance Time Excitation Functions from the first principles and building up toward a thorough application of the approach to seismic response analysis of structures and evaluation of the associated damages. The authors have also provided the best practice solution methods for practical application to aseismic building designs. Whether readers view this text from either the fundamentals of Endurance Time methodology or the earthquake engineering side, the book provides the readers with a clear understanding of the comprehensive treatment of the structural responses with the innovative approach of Endurance Time methodology for designing buildings against earthquake. This book is undoubtedly an enthusiastic celebration of three essential elements – theory, modeling, and practice – in handling a multitude of structural design issues under earthquake excitations.

Goodarz Ahmadi
Clarkson University
Potsdam
26 May 2021

Preface

The Endurance Time Method (ETM) is a relatively new approach to seismic assessment of structures in which intensifying dynamic excitations are used as the loading function. ETM has some unique advantages that make it distinctive among conventional seismic assessment procedures. First, it is a response history-based procedure and thus, inherits the unique capabilities of true dynamic procedures in the analysis of complex systems. Second, ETM reduces the computational effort needed in typical response history analysis by a factor roughly in the order of 10- to 100-fold. This advantage extends its application into the research areas where computational demand imposes a practical limitation in the application of conventional dynamic methods. The third important advantage of the method lies in its conceptual simplicity and uniform applicability to a wide range of earthquake engineering problems. These advantages make ETM a great tool for preliminary dynamic analysis of almost any structural system. Thus, practical applications of ETM in the field of earthquake engineering are expected to extensively develop in near future.

The essence of ETM is the application of intensifying dynamic excitations to the system of interest. These are called Endurance Time Excitation Functions, or shortly ETEFs. ETM is about applying ETEFs into a system and estimating its expected response when subjected to ground motions scaled to match seismic hazards of desired intensities of interest. Availability of appropriate ETEFs is critical in the quality, reliability, and accuracy of ET analysis results. Therefore, production of useful and efficient ETEFs is crucial in successful implementation and utilization of ETM.

This book presents a comprehensive source of information on various aspects of producing Endurance Time Excitation Functions. The book is a recompilation and combination of various research papers published on the subject by the authors in recent years. These source articles are selected and edited carefully in order to provide a coherent resource for researchers and practitioners who are interested in the topic. It is the first book on the topic of ETEFs, and it can be used for acquiring deeper knowledge and practical understanding of the core of ETM. Also, it can serve as a good starting point for those who are interested in producing ETEFs with desired characteristics and for particular applications.

We are indebted to Professor G. Ahmadi of the University of Clarkson for reviewing the manuscript of the book and offering his insightful feedback. We would like to thank Professor Mohammad Noori of the California Polytechnic State University for ongoing encouragement. Lastly, we would like to express our gratitude to Dr. Joseph Clements and Ms. Lisa Wilford from Taylor and Francis Group for their professionalism and support in the process of publication.

H. E. Estekanchi
M. Mashayekhi
A. Vafai

Authors

Homayoon E. Estekanchi is a Professor of Civil Engineering at Sharif University of Technology (SUT). He received his Ph.D. in Civil Engineering from SUT in 1997 and has been a faculty member at SUT since then. He is a member of Iranian Construction Engineers Organization, ASCE, Iranian Inventors Association, and several other professional associations. His research interests include a broad area of topics in structural and earthquake engineering with a special focus on the development of the Endurance Time Method and the Value-Based Seismic Design.

Mohammadreza Mashayekhi received his Ph.D. in Civil Engineering from Sharif University of Technology (SUT) in 2018. He is an Assistant Professor of Civil Engineering at K. N. Toosi University of Technology (KNTU), Iran. He has co-authored over 30 scientific papers and is a member of Iranian Construction Engineers Organization. He spent a postdoctoral year at SUT from 2019 to 2020. In recent years, he has been the executive editor of *Numerical Methods in Civil Engineering* journal, an international open-access, online, peer-review journal publishing original papers of high quality related to the numerical methods in all areas of civil engineering. His contributions are in Endurance Time Method, reliability-based analysis, and optimization in civil engineering.

Hassan A. Vafai has held the position of professorship in civil engineering at different universities including Sharif University of Technology (SUT), Washington State University, and University of Arizona. He was the Founder and Editor-in-Chief of *Scientia*, a peer-reviewed international journal of science and technology. Throughout his professional career, he has received numerous awards and distinctions including emeritus distinguished engineer by the National Academy of Sciences, Iran; an honorary doctorate by the Senatus Academicus of Moscow Region State Institution of Higher Education; and the "Order of Palm Academiques", awarded by the Ministry of Education, Research and Technology of France.

Nomenclature

A_i	Frequency amplitude
$a_{j,k}$	Discrete wavelet coefficient at the j-th time index and the k-th scale
$A_0(\omega)$	Fourier transform of signal
$a_0(t)$	Signal in time domain
a_g	Acceleration time history of endurance time excitations
B	Magnification factor
C	Earthquake coefficient
C_j	Normalized cost
country	Initial population in imperialist competitive algorithm
c_k	k-th wavelet coefficient
CAV	Cumulative absolute velocity
$\mathbf{CAV}_C$	Target cumulative absolute velocity
$CAV_{C,P}$	Target cumulative absolute velocity with exceedance probability P
dE_H	Incremental hysteretic energy demand
D_{PA}	Modified Park–Ang damage index
d	Distance between colony and imperialist
$\mathbf{D}_Y$	Diagonal matrix
EDP	Engineering demand parameter
$\mathbf{EDP}_P(t)$	Engineering demand parameter with exceedance probability P
$E_H^{\text{target}}(T,\mu)$	Target hysteretic energy of GMs
$E_H(t,T,\mu)$	Hysteretic energy of ETEFs at time t, period T, and ductility μ
$E_{HC}(t,T,\mu)$	Target hysteretic energy of simulating ETEFs at time t, period T, and ductility μ
Err_1	Error function 1
Err_2	Error function 2
$E(t,\omega)$	First error function
ET	Endurance Time
E_L	Modification of transformation coefficients
ETEF	Endurance Time Excitation Function
FEMA	Federal Emergency Management Agency
$F_{N,i}$	The normalized objective function value of simulated excitations based on the i-th initial point
$F_{\text{ETEF}}(a_g)$	Objective function of simulating ETEFs
F_P	Endurance time excitations objective function corresponding to the exceedance probability of P
$f_{\text{cost}}^{(imp,j)}$	Cost of the j-th imperialist
$\overline{F_N}$	Average simulated ETEFs
f_s	Restoring force
$f(\tau)$	Signal

$F_y(T,\mu)$	Minimum lateral strength capacity that a SDOF system with a period of T requires to avoid the average ductility ratio demands larger than μ
GA	Genetic Algorithm
g	Gradient of objective function
$g(t)$	Increasing profile
GM	Ground motion
H	Hessian matrix of objective function
$h_{EH}(t,T,\mu)$	Time-dependent variation function of target hysteretic energy demand
$h_u(t,T,\mu)$	Variation function of target nonlinear response in time
I	Importance vector
I	Importance factor
ICA	Imperialist Competitive Algorithm
IDA	Incremental Dynamic Analysis
IM	Intensity Measure
$IM_{ET}\,(edp)$	Ordinate of endurance time method curve
$IM_{IDA}\,(edp)$	Ordinate of incremental dynamic analysis curve
imS_a	Importance factors of acceleration spectra
imS_u	Importance factors of displacement spectra
imCAV	Importance factors of cumulative absolute velocity
INBC	Iranian National Building Code
k	Linear stiffness
$\mathbf{L_Y}$	Lower triangular decomposition of the correlation matrix R_{YY}
M	Level of considered discrete wavelet transform
m	Number of period points used for discretization of objective function
$\mathbf{M_Y}$	Mean vector
M_y	Yield moment capacity
N	Number of harmonic motion (Chapter 1)
N_{var}	Dimension of the optimization problem
n	Number of time points used for discretization of objective function
$N_{country}$	Number of all countries
NC_j	Number of colonies associated with the j-th empire
N_{col}	Number of colonies
NTC_j	Normalized total power of the j-th empire
n_E	Number of random initial points used in order to explore the sensitivity to initial points
n_f	Number of considered increasing sine functions
N_{imp}	Number of the most powerful countries (imperialists)
$NRes_{s_a}$	Normalized residual associated with acceleration spectra
$NRes_{E_H}$	Normalized residual associated with hysteretic energy
$NRes_{u_m}$	Normalized residual associated with nonlinear displacement
N_{rev}	Revolution number
n_{pop}	Population number
p_j	Probability of the j-th imperialist

p_{rev}	Revolution percentage of the countries
R	Strength factor or response reduction factor
r	Number of ductility points used for discretization of objective function
RP	Return period
PGV	Peak Ground Velocity
Res_{S_a}	Normalized residuals, respectively, associated with acceleration spectra
Res_{S_u}	Normalized residuals, respectively, associated with displacement spectra
$\textbf{Res}_{\text{CAV}}$	Normalized residuals, respectively, associated with CAV
$\textbf{Res}_T$	Total residual
$\mathbf{R}_{\text{YY}}$	Correlation coefficient matrix
$R(\omega)$	Multiplicative function
$R(t,\omega)$	Second error function
$\{V_1\}$	Unit vector that its start point is the previous location of the colony
s	Step size in optimization problem
s	Dispersion coefficient
$S_a^{\text{target}}(T)$	Target acceleration spectra
$S_a(t,T)$	Acceleration spectra of ETEFs at time t and period T
S_{ac}	Code spectrum
$S_{aC}(t,T)$	Target acceleration spectra of simulating ETEFs at time t and period T
$S_{aC,P}$	Target acceleration spectra with exceedance probability P
$S_{uC,P}$	Target displacement spectra with exceedance probability P
$S_{a,P}^{\text{target}}(T)$	Normalized ground motion targets acceleration spectra
$S_{u,P}^{\text{target}}(T)$	Normalized ground motion targets displacement spectra
$S_{CAV,P}^{\text{target}}(T)$	Normalized ground motion targets cumulative absolute velocity
SDOF	Single degree of freedom
T	Period
TRR	Total relative residual
TRC	Total relative cost
TC_j	Total cost of the j-th empire
t	Time
t_{target}	Target time
$T_{\max}$	Maximum considered period in simulating endurance time excitations
$t_{\max}$	Duration of endurance time excitations
$u_m(t,T,\mu)$	Nonlinear displacement of ETEFs at time t, period T, and ductility μ
$u_{mC}(t,T,\mu)$	Target nonlinear displacement of ETEFs at time t, period T, and ductility μ
$u(\tau)$	Displacement response of nonlinear SDOF subjected to ETEF
$\dot{u}(\tau)$	Velocity response of nonlinear SDOF subjected to ETEF
$\ddot{u}(\tau)$	Relative acceleration response of SDOF subjected to ETEF
$\ddot{x}(\tau)$	Relative acceleration response of a single degree of freedom at time τ

x_i	i-th decision variable in optimization problem
$x_{i,\max}$	Upper bound of optimization variables
$x_{i,\min}$	Lower bound of optimization variables
α	Weight constant (Chapter 7)
α	Profile intensifying rate parameters (Chapter 2)
α_{S_u}	Weight factor of displacement in objective function
α_{CAV}	Weight factor of cumulative absolute velocity in objective function
α_{E_H}	Weight factor of hysteretic energy in objective function
α_{u_m}	Weight factor of nonlinear displacement in objective function
β	Amplitude movement parameter (in Chapter 7), Profile intensifying rate parameters (in Chapter 6, Profile intensifying rate parameters (in Chapter 2)
γ	Profile intensifying rate parameters
Δ	Spherical radius
ζ	A positive number indicates larger influence of the mean power of colonies in determining the total power of the empire
θ_m	Member-end rotation
θ_r	Recoverable rotation
θ_u	Ultimate rotation
λ	Scale factor
μ	Ductility ratio
$\mu_{\ln S_a(T)}$	Median acceleration spectra
$\mu_{\max}$	Maximum considered ductility ration in simulating ETEFs
μ_{rev}	Percentage of colonies which are chosen for revolution
$\mu_{\ln IM\vert EDP}$	Median intensity measure IM given engineering demand parameter EDP
ζ	Damping ratio
σ_{rev}	Standard deviation of revolution
Σ_{YY}	Covariance matrix
τ	Time
Φ	Cumulative normal distribution
ϕ	Scaling function
ϕ_i	Phase of motion at the i-th frequency
ψ	Wavelet function
Ω	Maximum absolute operator
ω	Angular frequency
$\omega_{\min}$	Minimum considered frequencies
$\omega_{\max}$	Maximum considered frequencies
ω_L^i	Lower frequency limits of the i-th frequency band
ω_U^i	Upper frequency limits of the i-th frequency band

1 Introduction to the Endurance Time Method and Intensifying Dynamic Load Functions

REVIEW

In this chapter, the basic concepts of the endurance time method along with a brief history of its development are presented. Diverse applications of this method in earthquake engineering are briefly explained. Afterward, Endurance Time Excitation Functions (ETEFs) as the core component of the ET method are introduced. In this regard, basic characteristics and different types of ETEFs are explained.

CONCEPT OF ENDURANCE TIME METHOD

Various natural hazards including earthquake threaten communities. A multitude of earthquakes (e.g., Tabas in 1978, Morgan Hill in 1984, Loma Prieta in 1989, Northridge in 1994, Bam in 2003) have struck cities all around the world and resulted in high casualty and monetary losses. Realistic anticipation of infrastructure's response to these hazards is necessary to achieve sustainable living environment. Several design and rehabilitation codes are developed to provide provisions for seismic assessment of structures. Seismic rehabilitation provisions (e.g., ASCE/SEI 41-17 and FEMA356) generally present four procedures for seismic analysis: Linear Static Procedure, Linear Dynamic Procedure, Nonlinear Static Procedure, and Nonlinear Dynamic Procedure. Among these procedures, nonlinear dynamic procedure considers both nonlinearity of materials and the dynamic nature of earthquakes, and it is thus taken to be the most reliable one. Nevertheless, there are several problems in the face of conventional nonlinear dynamic procedure methods such as incremental dynamic analysis, alias IDA (Vamvatsikos and Cornell, 2002). Of these are challenges of record selection (Araújo et al., 2016; Baker and Cornell, 2008) and huge computational demand of dynamic analysis, which prove serious hindrance to widespread application of nonlinear dynamic procedure in professional practice.

Several attempts were made by a number of researchers in the field to address the aforementioned difficulties. In this context, Endurance Time Analysis has been developed to function as a simple yet efficient tool for time history dynamic analysis of structural systems. Inspired by the exercise test in medicine (Riahi et al., 2009), the concept of the endurance time method was established by Estekanchi et al. (2004). The concept of ET method is somewhat similar to the exercise test conducted in medicine

DOI: 10.1201/9781003216681-1

in order to assess the cardiovascular condition of athletes or heart patients. In exercise test, the subject is asked to walk on a treadmill while its slope and speed is gradually increased. The test is continued until the subject is exhausted or abnormal biological conditions are observed. Based on the maximum speed and slope that the subject has been capable of enduring, his or her cardiovascular condition is assessed. Endurance time method follows the similar concept applied to structures. In the endurance time method, structures are subjected to a set of intensifying acceleration functions and response of structures is monitored as the analysis time is passed.

The concept of endurance time method can be explained by a hypothetical test. In this example, the objective is to determine the performance of three structures in earthquake. These structures are located in a shaking table as shown in Figure 1.1. These structures are subjected to a random excitation, amplitude of which increases with time. With the intensification of the amplitude of excitation, vibration amplitude of the structures increases, and hence, the demand of the structures increases. As the time passes in this test, structures gradually move from elastic region to nonlinear and they finally collapse. Damage indicators of these structures are monitored during the test. For example, maximum interstory drift ratio can be plotted against time.

As can be seen in Figure 1.1, the extent of damage in structures A and B is, respectively, the maximum and the minimum among the considered structures. Therefore, it can be concluded that the performance of structure B is more desirable than two other structures at different intensities. Another issue is the damage capacity of structures. Figure 1.2 shows that structures A, B, and C collapse at 8, 13, and 18 seconds, respectively. By correlating this time with corresponding intensity of the excitation at the collapse time, the endurance of each structure can be linked to the maximum time that the structure can endure the intensifying excitation. The main idea here

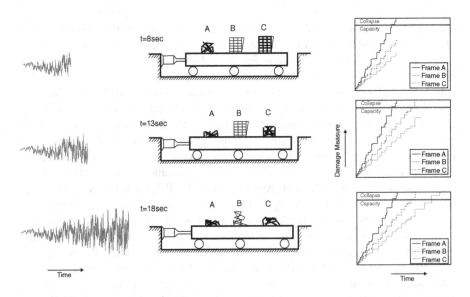

FIGURE 1.1 The concept of the endurance time method for determining the seismic demand of three structures.

is to correlate the intensity of the input excitation to that of real ground motions. If these three structures are designed for a same seismicity level, structure B is the best design in terms of strength and endurance. If input excitations are calibrated such that they fulfill the code design spectra requirements, a minimum acceptable endurance time can be specified for practical assessment. In fact, the performance of structures can be quantified by the endurance time. The interesting issue is that the above deductions are merely based on observations of structural dynamic behavior and there is no need for knowledge of dynamic characteristics of the structures in the interpretation of the observation. Of course, these dynamic characteristics influence the endurance time, but each structure performance can be judged based on its observed endurance.

The increasing trend of increasing load functions introduces new interpretation of time in this analysis. Time in the endurance time analysis corresponds to the intensity of the excitation. For example, in Figure 1.2, the maximum base shear is plotted against time. As it can be seen, intensity measure increases with time. Endurance time analysis results are presented by increasing curves in which horizontal axis is time as an indicator of intensity and vertical axis is a response parameter of the structure that is the maximum response of the structure up to time t. Vertical axis values can be formulated as follows:

$$\Omega(f(t)) = Max(\|f(\tau)\|) \quad 0 \leq \tau \leq t \tag{1.1}$$

In the above equation, Ω is the maximum response in the time span $[0,t]$ and f is the response history as a function of time. Maximum drift, base shear, and plastic

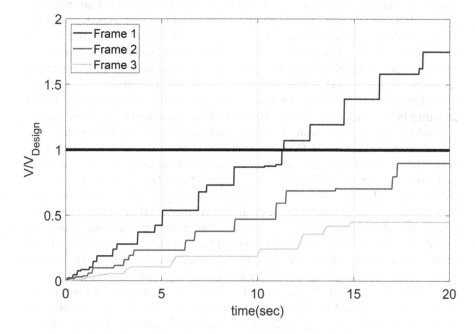

FIGURE 1.2   Increasing response plot of three structures: A, B, and C.

rotation can be considered as response. In view of Equation (1.1), if the maximum interstory drift is taken as response parameter, Ω is the maximum drift ratio that the structure experienced in up to time t.

At early, mid-, and late stages of endurance time excitations, load intensities are low, moderate, and large, and therefore, if properly calibrated, excitations at these stages can be considered to be representative of low, moderate, and huge earthquakes, respectively. In fact, in endurance time method, time is an intensity indicator. The intensification of endurance time excitations leads to inclusion of a wide range of intensity levels in a single time history analysis. By correlating time and intensity measure, the maximum seismic demand on structures for various intensity levels can be determined as a function of intensity measure. Therefore, in the endurance time method, structural responses at different intensity levels are obtained in a single time history analysis, thereby reducing the computational demand. This is the main difference between the endurance time method and conventional time history analyses which require separate response history analyses at each intensity level.

APPLICATIONS OF ENDURANCE TIME METHOD

Endurance time method is currently used in different areas by earthquake engineers. Applications of this novel method are expanding. Estekanchi et al. (2007) used endurance time method in linear dynamic analysis. They first introduced a procedure to generate endurance time load functions consistent with a design code spectrum. They compared linear dynamic analysis results of braced steel structures obtained by endurance time method with that obtained by spectrum dynamic analysis. It is shown that code-based endurance time method can estimate interstory drift ratio and internal forces of equivalent static analysis and spectrum dynamic analysis with acceptable accuracy. They showed that equivalent static analysis results for irregular structures diverge from spectrum dynamic analysis results, while endurance time method estimation results are acceptable for irregular structures. Riahi et al. (2009) investigated the potential of endurance time method in nonlinear dynamic analysis of structures. In their study, single degree-of-freedom systems with different periods, ductility ratios, and different degrading behaviors are considered. They showed that endurance time method leads to acceptable correspondence with time history analysis. Rahimi and Estekanchi (2015) used endurance time method for determining the collapse assessment of steel structures. Riahi et al. (2010) employed endurance time method for seismic assessment of steel structures with different stories and different bays. They considered interstory drift ratio and plastic rotation. Results demonstrate the good performance of endurance time analysis in seismic response prediction of time history analysis. Mirzaee and Estekanchi (2015) proposed a new methodology based on endurance time analysis for retrofitting the existing structures. They applied the new methodology in several steel structures and obtained the optimum retrofitting procedure. Chiniforush et al. (2016) applied the ET method in seismic evaluation of unreinforced masonry monuments. Basim and Estekanchi (2015) applied the ET method in performance-based optimum design. Valamanesh and Estekanchi (2013) adopted this method for multicomponent analysis of structures. Riahi et al. (2015) assessed seismic collapse of reinforced concrete moment frames using the endurance time method. They considered 30 concrete structures with degrading behavior. Results obtained by endurance time method are compared with those obtained by incremental dynamic analysis. Comparisons show that acceleration spectrum in which structures collapse in endurance time analysis is consistent with that in incremental dynamic analysis. Another interesting point they mentioned was that endurance time method well predicts collapse pattern for the considered structures especially for tall buildings. Estekanchi et al. (2018) used the ET method to investigate the interaction of moment-resisting frames and shear walls in reinforced concrete dual systems. Basim et al. (2016) proposed a methodology for value-based design by using endurance time method. Alembagheri and Estekanchi (2011) utilized endurance time method for seismic assessment of anchored and unanchored concrete tanks by using endurance time method. Maleki-amin and Estekanchi (2017) employed endurance time method for damage assessment of steel structures. They used damage spectrum to determine target time. Hariri-Ardebili et al. (2014) used endurance time method for performance-based seismic assessment of steel structures. They also evaluated progressive damage of steel structures by using endurance time method. They showed that endurance time method well predicts time history analysis results with considerably low computation time with acceptable accuracy. Hasani et al. (2017) employed endurance time method in nonlinear seismic analysis of offshore pile-supported systems. They applied endurance time method to a functional offshore jacket platform in the Persian Gulf region. Seismic response of the considered models by endurance time method has been compared with conventional time history method. The results indicate that endurance time method is reliable in capturing seismic response of offshore platforms supported on piles with an acceptable accuracy.

IMPLEMENTATION OF THE ENDURANCE TIME METHOD

As explained before, the endurance time method is a time history dynamic analysis in which structures are subjected to intensifying artificial acceleration functions. This method reduces complexity and computational demand of conventional nonlinear seismic analysis, and it provides response at different seismic levels in a single time history analysis.

Three factors play major roles in the accuracy of endurance time method: intensifying load functions, endurance criteria, and structural model. Among the mentioned factors, intensifying load function is the core of the method. In order to achieve useful and reliable results corresponding to real ground motions, intensifying load functions must be appropriately calibrated. Determination of the most appropriate and optimal dynamic input function, in itself, is the subject of extensive research work and is covered in this book.

INTENSIFYING DYNAMIC LOAD FUNCTIONS

Practical use of endurance time method calls for appropriate endurance time excitations that lead to consistent analysis results with time history analysis. There are various versions of endurance time excitations with different lengths, characteristics, and optimization techniques used in their development.

Simulating endurance time excitations usually involves a trial-and-error procedure. In fact, endurance time excitations are first generated and then are employed in dynamic analysis. If satisfactory results by using simulated excitations are not obtained, the procedure is modified and new excitations are simulated. This iterative loop is continued until satisfactory dynamic results are achieved. Since ground motions in each region depend on geotechnical characteristics, acceleration functions may need to be separately generated for each region.

The main characteristic of endurance time excitations is that the load and response of structures subjected to them constantly increase with time. This feature of endurance time excitations differentiates them from ground motions. The amplitude increase profile can take various forms. In initial endurance time excitations, linear profile in which maximum acceleration is directly proportional to time was adopted as the simplest practical form. Three conceptually different profiles are shown in Figure 1.3.

Real ground motions are usually defined by three main characteristics: intensity, frequency content, and duration. In order to simulate efficient endurance time excitations, these three features have to be considered. Frequency content and amplitude depend on earthquake magnitude, distance of the site from source, and soil condition, and they differ from one ground motion to another. Selection of target frequency content for simulating endurance time excitations is a challenging problem. Since frequency content and amplitude of motions are reflected in response spectra, considering response spectra in the simulation process implies that amplitude and frequency content are suitably considered. Two excitations with the same acceleration spectra induce equal responses in a structure. Almost all design codes define earthquake loads based on acceleration spectra.

In the development of endurance time method, five generations of endurance time excitations have been introduced so far. These endurance time excitation generations differ in terms of the simulation procedure and considered dynamic characteristics. The simulation of the first generation was based on random vibration theory, while other four generations were simulated based on optimization techniques. The second, third, fourth, and fifth generations differ in terms of considered dynamic characteristics in the simulation procedure. In the simulation of the second generation, linear response spectra are considered. Nonlinear response spectra are included in the simulation of the third generation. Duration consistency is added for the production

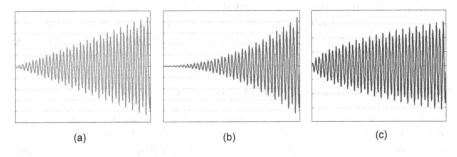

(a)   (b)   (c)

FIGURE 1.3   Different increasing profiles: (a) linear, (b) ascending rate, and (c) descending rate.

of the fourth generation. The main advantage of the fifth generation over other generations is damage consistency.

The first generation is only generated to introduce the basic concept of endurance time method. This generation was not employed for practical use. In this chapter, the first generation is explained to introduce the endurance time method concept and other four generations will be explained in the next chapters.

In the simulation of the first generation, the concept of nonstationary random pattern in amplitude and frequency content is used. In this method, excitation functions are expressed by summing a finite number of harmonic functions as follows:

$$a(t) = \sum_{i=1}^{N} A_i \sin(\omega_i t + \phi_i) \qquad (1.2)$$

where N is the number of harmonic functions, ω_i and φ_i are the angular frequency and phase angle of i-th sine function, and A_i is the frequency amplitude. If amplitudes of these sine functions are kept constant and the phase angles are changed, different motions with different frequency contents are generated. Excitations that are generated by this method are called stationary excitations.

Random excitation with Gaussian distribution with mean 0 and standard deviation can be used as a starting point. The number of acceleration data points of 2048 was chosen. The maximum acceleration of 1.0g was assumed. In order to adjust the signal to appropriate scale, an appropriate modulating function was adopted. This modulating function was selected based on earthquake intensity and distance for the earthquake source. In order to remove undesirable frequency range, the excitations were filtered by inverse Fourier transform technique. The objective was to simulate an excitation that matches a particular spectrum. The target spectrum was the code spectrum of standard 2800 Iranian Seismic Code for soil type II. If the acceleration spectra of the motion were different with the target spectrum, the ratio of target spectrum to motion spectra was multiplied in the Fourier amplitude of motion. This procedure was repeated until the acceptable spectral match was achieved. The resulting stationary accelerogram was then modified by applying a linear profile function that made it intensifying, with respect to peak accelerations, at different time intervals. These accelerograms served well for the purpose of demonstrating concept of endurance time analysis, but could not be expected to result in quantitatively reliable results. A sample of first-generation simulated excitations is depicted in Figure 1.4. As it can be seen, the intensity of excitations increases with time.

SIMULATED ENDURANCE TIME EXCITATIONS

Specifications of the existing simulated endurance time excitations are presented in this section. Acceleration load functions differ in terms of the target spectrum, periods range considered in optimization process, objective functions, time step, duration length, and optimization algorithm. As mentioned in the previous section, the first generation of ETEFs was only used in order to demonstrate the concept of ET method and is not intended for practical application. The second-generation ETEFs mentioned

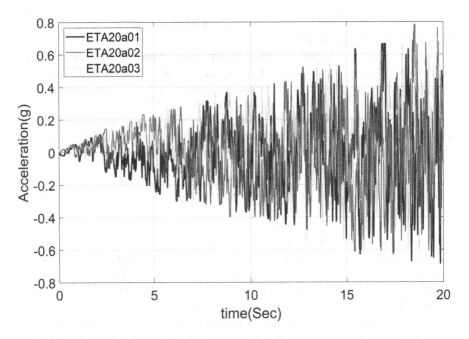

FIGURE 1.4 First-generation endurance time excitation.

in Table 1.1 were produced by fitting their response spectra to a design code spectra. Series ETA20a and ETA20b belong to this category. The periodic range that is covered in their optimization is from 0 to 5 seconds. These records produce reasonably good results in the linear systems but are not suitable for nonlinear analysis. Periodic range considered in the optimization process has appreciable impact on nonlinear charac- teristics of endurance time excitations. By including long periods in the simulation of second-generation excitations, they can be used in nonlinear dynamic analysis. ETEF series ETA20e and ETA20f correspond to this category. Nonlinear spectra can be

TABLE 1.1

Properties of Some ETEFs

ETEF Series	Generation	Optimization Scope	Applicability	Template Spectrum	Notes
ETA20a	2	Linear	Linear	2800 v3 type II	Outdated. Useful as optimization benchmarks
ETA20b	2	Linear	Linear	2800 v3 type II	Same as ETA20a series
ETA20e	2	Linear	Nonlinear	FEMA 440 C 7 GMs	Outdated
ETA20en	3	Nonlinear	Nonlinear	FEMA 440 C 7 GMs	Superseded by ETA20in series
ETA20f	2	Linear	Nonlinear	FEMA 440 C 7 GMs	Outdated

(Continued)

TABLE 1.1 (*Continued*)
Properties of Some ETEFs

ETEF Series	Generation	Optimization Scope	Applicability	Template Spectrum	Notes
ETA20inx ETA20iny ETA20inz	3	Nonlinear	Nonlinear	FEMA 440 C 20 GMs	Recommended for general and nonlinear analysis Multicomponent
ETA20jn	3	Nonlinear	Nonlinear	ASCE07	Good for design applications
ETA40g	2	Linear	Nonlinear	ASCE07	Covers higher intensities up to a scale factor of 4
ETA40h	2	Linear	Nonlinear	FEMA 440 C 7 GMs	Covers higher intensities up to a scale factor of 4
ETA40lc	4	Duration	Nonlinear	FEMA p695 FF	Good for nonlinear analysis. CAV consistent with scaled-up GM set
ETA20kd	5	Damage	Nonlinear	FEMA p695 FF	Good for nonlinear analysis. Better damage consistency achieved

included in the process of generating ETEFs that perform well in nonlinear analysis. These constitute the third generation of ETEFs. ETEF series ETA20en and ETA20in series. The ETA20in series is produced for x, y, and z (vertical) spectra of FEMA 440 soil type C records. So, these can be used for multicomponent analysis as well. In the fourth generation of the ETEFs, duration consistency is also considered. Series ETA40lc belongs to this generation. In this series, CAV has been considered as a duration indicator and implemented in their generation. In the fifth generation, damage consistency with scaled ground motions is also included in the generation of intensifying loads. Series ETA20kd belongs to the fifth generation. The specifications of some publicly available ETEFs are summarized in Table 1.1. These excitations are available on ET website (H.E. Estekanchi, 2016). Details on challenges of producing ETEFs will be explained in the next chapters of this book.

REFERENCES

Alembagheri, M., & Estekanchi, H. (2011). Seismic analysis of steel liquid storage tanks by endurance time method. *Earthquake Engineering and Engineering Vibration*, *10*, 591–604. https://doi.org/10.1016/j.tws.2011.08.015

Araújo, M., Macedo, L., Mário, M., & Castro, J. (2016). Code-based record selection methods for seismic performance assessment of buildings. *Earthquake Engineering and Structural Dynamics*, *45*, 129–148. https://doi.org/10.1002/eqe

ASCE/SEI 41-17. (2017). *Seismic Evaluation and Retrofit of Existing Buildings (41-17)*. Reston, VA: American Society of Civil Engineers.

Baker, J. W., & Cornell, C. A. (2008). Vector-valued intensity measures for pulse-like near-fault ground motions. *Engineering Structures*, *30*(4), 1048–1057. https://doi.org/10.1016/j.engstruct.2007.07.009

Basim, M. C., & Estekanchi, H. E. (2015). Application of endurance time method in performance-based optimum design of structures. *Structural Safety*, *56*, 52–67. https://doi.org/10.1016/j.strusafe.2015.05.005

Basim, M. C., Estekanchi, H. E., & Vafai, A. (2016). A methodology for value based seismic design of structures. *Scientia Iranica*, *23*(6), 2514–2527.

Chiniforush, A., Estekanchi, H.., & Dolatshahi, K. (2016). Application of endurance time analysis in seismic evaluation of an unreinforced masonry monument. *Journal of Earthquake Engineering*, *23*(3), 827–841. https://doi.org/10.1080/13632469.2016.1160008

Estekanchi, H. E. (2016). Endurance Time Method Website. Estekanchi, H. (2016). Endurance Time Method. Available online: https://sites.google.com/site/etmethod/.

Estekanchi, H. E., Harati, M., & Mashayekhi, M. (2018). An investigation on the interaction of moment-resisting frames and shear walls in RC dual systems using endurance time method. *The Structural Design of Tall and Special Buildings*, *27*(12), 1–16. https://doi.org/10.1002/tal.1489

Estekanchi, H. E., Vafai, A., & Sadeghazar, M. (2004). Endurance time method for seismic analysis and design of structures. *Scientia Iranica*, *11*(4), 361–370.

Estekanchi, H. E., Valamanesh, V., & Vafai, A. (2007). Application of endurance time method in linear seismic analysis. *Engineering Structures*, *29*(10), 2551–2562. https://doi.org/10.1016/j.engstruct.2007.01.009

FEMA356. (2000). *NEHRP Guidelines for the Seismic Rehabilitation of Buildings, FEMA 356 (Misc)*. Washington, DC: Federal Emergency Management Agency.

Hariri-Ardebili, M. A., Sattar, S., & Estekanchi, H. E. (2014). Performance-based seismic assessment of steel frames using endurance time analysis. *Engineering Structures*, *69*, 216–234. https://doi.org/10.1016/j.engstruct.2014.03.019

Hasani, H., Golafshani, A. A., & Estekanchi, H. E. (2017). Seismic performance evaluation of jacket-type offshore platforms using endurance time method considering soil-pile-superstructure interaction. *Scientia Iranica*, *24*(4), 1843–1854. https://doi.org/10.24200/sci.2017.4275

Maleki-amin, M. J., & Estekanchi, H. E. (2017). Damage estimation of steel moment-resisting frames by endurance time method using damage-based target time. *Journal of Earthquake Engineering*, *2469*, 1–30. https://doi.org/10.1080/13632469.2017.1297265

Mirzaee, A., & Estekanchi, H. E. (2015). Performance-based seismic retrofitting of steel frames by the endurance time method. *Earthquake Spectra*, *31*(1), 383–402. https://doi.org/10.1193/081312EQS262M

Rahimi, E., & Estekanchi, H. E. (2015). Collapse assessment of steel moment frames using endurance time method. *Earthquake Engineering and Engineering Vibration*, *14*(2), 347–360. https://doi.org/10.1007/s11803-015-0027-0

Riahi, H. T., Amouzegar, H., & Falsafioun, M. (2015). Seismic collapse assessment of reinforced concrete moment frames using endurance time analysis. *The Structural Design of Tall and Special Buildings*, *24*(4), 300–315. https://doi.org/10.1002/tal

Riahi, H. T., Estekanchi, H. E., & Vafai, A. (2009). Application of endurance time method in nonlinear seismic analysis of SDOF systems. *Journal of Applied Sciences*, *9*(10), 1817–1832.

Riahi, H. T., Estekanchi, H. E., & Vafai, A. (2010). Seismic assessment of steel frames with the endurance time method. *Journal of Constructional Steel Research*, *66*(6), 780–792. https://doi.org/10.1016/j.jcsr.2009.12.001

Valamanesh, V., & Estekanchi, H. E. (2013). Compatibility of the endurance time method with codified seismic analysis approaches on three-dimensional analysis of steel frames. *The Structural Design of Tall and Special Buildings*, *22*(2), 144–164.

Vamvatsikos, D., & Cornell, C. A. (2002). Incremental dynamic analysis. *Earthquake Engineering and Structural Dynamics*, *31*(3), 491–514. https://doi.org/10.1002/eqe.141

2 Objective Functions for Generating Endurance Time Excitation Functions[1]

REVIEW

In this chapter, the objective functions used for simulating endurance time excitations associated with the second, third, fourth, and fifth generations are explained. Dynamic characteristics included in simulating each generation are described. Calculation of target and endurance time excitation dynamic characteristics is also explained. As expressed in Chapter 1, simulation of all generations of ETEFs except for the first generation is based on optimization processes. Objective functions are in the core of formulating any optimization problem, as explained in this section.

SECOND GENERATION OF ENDURANCE TIME EXCITATIONS

Response spectra concept was adopted to simulate second-generation endurance time excitations. It was assumed that if acceleration spectra of an excitation match with the acceleration spectrum of a ground motion, the maximum response of endurance time method and time history analysis will be consistent. In fact, in the second generation of endurance time excitations, the concept of response spectra is directly involved. Target spectrum can be average spectrum of a suite of ground motions or a code spectrum.

Second-generation endurance time excitations are designed in such a way as to produce dynamic responses equal to the code design spectrum at a predefined time, t_{target}, and therefore, it is possible to compare the performance of different structures with different dynamic characteristics. The choice of an appropriate target time should also be given consideration. In the general nonlinear case, if the target time is too low, the structure may not have time to go through a reasonable number of nonlinear cycles. On the other hand, if the target time is too high, the number of cycles will become unrealistically high. In the linear range, however, the chosen

---

[1] Chapter Source:
  1. Mashayekhi, M., Estekanchi, H. E., Vafai, H., & Mirfarhadi, S. A. (2018). Development of hysteretic energy compatible endurance time excitations and its application. *Engineering Structures*, *177*, 753–769.
  2. Mashayekhi, M., Estekanchi, H. E., Vafai, A., & Mirfarhadi, S. A. (2018) Simulation of cumulative absolute velocity consistent endurance time excitations. *Journal of Earthquake Engineering*, *25*(5), 892–917.

DOI: 10.1201/9781003216681-2

value of the target time is rather arbitrary and does not have any significant effect on the results, i.e., by scaling the accelerograms using linear scale factor, S_a or S_d can be set to reach the required level at any desired time. The target value of 10 seconds has been used for scaling of the accelerograms. It can be anticipated that the optimal value of target time should somehow be correlated to natural period of vibration of the structure and also on the duration of representative earthquakes.

Similar to the first generation, the linear intensification scheme is maintained until improved scheme can be suggested through further research. In this scheme, response spectrum of an endurance time excitation is to increase proportionally with time, i.e., at time $t_{target} = t_{target}/2$, the response is to be half of the codified value; at $t_{target} = 2t_{target}$, it is to be twice the codified value; etc. In this way, the target acceleration response spectra of endurance time excitation are defined as follows:

$$S_{aC}(T,t) = \frac{t}{t_{target}} \times S_{aC}(T) \tag{2.1}$$

where $S_{aC}(T,t)$ is the target acceleration response spectra at time t and period T. $S_{sC}(T)$ is the target acceleration spectrum which can be codified design acceleration spectrum. Displacement response spectrum is also a significant parameter that is closely related to the acceleration spectra. The target displacement response spectra can be defined as follows:

$$S_{uC}(T,t) = \frac{t}{t_{target}} \times S_{aC}(T) \times \frac{T^2}{4\pi^2} \tag{2.2}$$

where $S_{uC}(T,t)$ is the target displacement response spectra at time t and period T.

Acceleration and displacement spectra of endurance time excitations are computed according to the following equations:

$$S_u(T,t) = \max\left(\left|u(T,\tau)\right|\right) \quad \tau \in [0,t] \tag{2.3}$$

$$S_a(T,t) = \max\left(\left|a(T,\tau)\right|\right) \quad \tau \in [0,t] \tag{2.4}$$

where $u(T,\tau)$ and $a(T,\tau)$ are the displacement and acceleration response history of a single degree of freedom with period T at time τ. $S_u(T,\tau)$ and $S_a(T,\tau)$ are displacement and acceleration response spectra of endurance time excitation at period T and time t.

Analytical approaches to find an accelerogram that satisfies the target response defined by Equations (2.1) and (2.2) are formidably complicated. Therefore, the problem was approached by formulating it as an unconstrained optimization problem in the time domain as follows:

$$\min F(a_g) = \int_0^{T_{max}} \int_0^{t_{max}} \left\{ \left[S_a(T,t) - S_{aC}(T,t)\right]^2 + \alpha\left[S_u(T,t) - S_{uC}(T,t)\right]^2 \right\} dt\, dT \tag{2.5}$$

where a_g is the endurance time excitation being sought and α is the relative weight parameter that can adjust the effective penalty attributed to displacement deviation as compared to acceleration deviations from target values. High values of α make the target function more sensitive to displacement response differences, whereas low values of α make the target function more sensitive to acceleration spectra differences. Considering the theoretical correlation between acceleration response and relative displacement response, a high value of α results in a better fit in the high period range and a low value of α results in a better fit in the low period range. In the simulation of second-generation endurance time excitations, the value of α has been set to 1.0, forcing a balanced penalty on either response parameter.

THIRD GENERATION OF ENDURANCE TIME EXCITATIONS

In the simulation of the third generation, compatibility of nonlinear responses is also considered. Similar to the formulation of simulating second-generation endurance time excitations, simulating the third generation can be defined by using the strength factor R as formulated below:

$$S_u(T,t,R) = \max\left(\left|u(\tau)\right|\right) \; \tau \in [0,t] \tag{2.6}$$

$$S_{uT}(T,t,R) = \frac{t}{t_{target}} \times S_{uT}(T,R) \times \frac{T^2}{4\pi^2} \; \tau \in [0,t] \tag{2.7}$$

$$\min F(a_g) = \int_0^{T_{max}} \int_0^{R_{max}} \int_0^{t_{max}} \left[S_u(T,t,R) - S_{uT}(T,t,R)\right]^2 dt\,dR\,dT \tag{2.8}$$

In this formulation, $R = 1$ corresponds to the linear model, so a rather simple objective function such as Equation (2.8) can be used for producing third-generation ETEFs. Ductility factor (μ) can also be used for including nonlinearity in the optimization process. It should be noted that producing third-generation ETEFs requires hugely more computational demand as the nonlinear analysis involved in the optimization process. The details of this process will be explained in the next sections.

FOURTH GENERATION OF ENDURANCE TIME EXCITATIONS

In the fourth generation, duration-related parameters are also included in simulation procedure. Although acceleration spectra are good representatives of intensity and frequency contents of ground motions, however, in order to quantify ground motions, it is required to consider not only the intensity and frequency content but also the duration characteristics. There is a body of research to back this notion. Hancock and Bommer (2006) reviewed papers which had investigated the influence of strong motion duration on structural responses. They state that there are several hundred papers which relate structural damage to the duration of ground motions, either directly or indirectly, and conclude that there is positive correlation between damage related to cumulative energy and strong motion duration.

Therefore, duration must be included in the generating process, meaning that it must first be correctly quantified. A variety of definitions have been proposed for duration in the literature of earthquake engineering (e.g., Electrical Power Research Institute, 1991; Wang et al., 2016; Bommer and Martinez-Periera, 1999; Bolt, 1973). There are three types of strong motion duration definitions, i.e., bracketed duration that defines strong motion duration as the time interval between the first and the last exceedances of a particular threshold (e.g., 0.05 g), (b) uniform duration which is defined as the sum of intervals during which the record exceeds a specified acceleration threshold (e.g., 0.05 g), and (c) effective duration defined as time interval between two particular thresholds of the Arias intensity. Bommer and Martinez-Periera (1999) reviewed more than 30 different strong motion duration definitions. Although these definitions are widely used for actual ground motions, they cannot be directly applicable in simulating ETEFs.

Another way to consider the duration effect is to determine an intensity measure capable of describing the damage potential of ground motions. In order to generate better correlation between earthquake energy content and structural damage, alternative ground motion parameters have been proposed. The main purpose has been to decrease dispersion in predicted values associated with the ground motions, and it has been observed that the seismic performance of structures does correspond to the integration of acceleration time histories accordingly. Arias intensity (AI) and cumulative absolute velocity (CAV) have therefore been defined based on the integration of acceleration. Campbell and Bozorgnia (2012) suggest that CAV might be used to rapidly assess potential damage to a general class of conventional structures. CAV is mathematically represented by the following equation:

$$\text{CAV} = \int_0^{t_{\max}} |a(t)| \, dt \tag{2.9}$$

where $a(t)$ is the acceleration time history of GMs and $t_{\max}$ is the duration.

The ultimate goal is to generate endurance time excitations that appear in total correspondence with recorded ground motions. In order to fulfill this expectation, the characteristics of endurance time excitations must be compatible with the recorded ground motions. The method of generating endurance time excitations is to minimize the discrepancy between endurance time excitations and ground motions. Objective functions of Equations (2.10)–(2.12) compute residuals in three different ways: absolute residuals, relative residuals, and a combination of absolute and relative residuals, respectively. These objective functions integrate residuals over all times and periods. It should be noted that acceleration spectra, linear displacement spectra, and cumulative absolute velocity are considered in these objective functions:

$$F_{\text{CAV-A}}(a_g) = \int_{T_{\min}}^{T_{\max}} \int_0^{t_{\max}} \left\{ \begin{array}{l} \left[S_a(T,t) - S_{aC}(T,t) \right]^2 + \\ \alpha_{S_u} \left[S_u(T,t) - S_{uC}(T,t) \right]^2 \\ + \alpha_{CAV} \left[CAV(t) - CAV_C(t) \right]^2 \end{array} \right\} dt \, dT \tag{2.10}$$

$$F_{\text{CAV-R}}\left(a_g\right) = \int_{T_{\min}}^{T_{\max}} \int_0^{t_{\max}} \left\{ \begin{bmatrix} \dfrac{S_a\left(T,t\right) - S_{aC}\left(T,t\right)}{S_{aC}\left(T,t\right)} \end{bmatrix}^2 + \\ \begin{bmatrix} \dfrac{S_u\left(T,t\right) - S_{uC}\left(T,t\right)}{S_{uC}\left(T,t\right)} \end{bmatrix}^2 \\ \begin{bmatrix} \dfrac{CAV\left(t\right) - CAV_C\left(t\right)}{CAV_C\left(t\right)} \end{bmatrix}^2 \end{bmatrix} \right\} dt\, dT \tag{2.11}$$

$$F_{\text{CAV-M}}\left(a_g\right) = \int_{T_{\min}}^{T_{\max}} \int_0^{t_{\max}} \left\{ \begin{bmatrix} \dfrac{S_a\left(T,t\right) - S_{aC}\left(T,t\right)}{S_{aC}\left(T,t\right)} \end{bmatrix}^2 + \\ \begin{bmatrix} \dfrac{S_u\left(T,t\right) - S_{uC}\left(T,t\right)}{S_{uC}\left(T,t\right)} \end{bmatrix}^2 \\ \begin{bmatrix} \dfrac{CAV\left(t\right) - CAV_C\left(t\right)}{CAV_C\left(t\right)} \end{bmatrix}^2 + \\ \left[S_a\left(T,t\right) - S_{aC}\left(T,t\right) \right]^2 \\ + \alpha_{S_u}\left[S_u\left(T,t\right) - S_{uC}\left(T,t\right) \right]^2 + \\ \alpha_{CAV}\left[CAV\left(t\right) - CAV_C\left(t\right) \right]^2 \end{bmatrix} \right\} dt\, dT \tag{2.12}$$

where a_g is the acceleration time history of ETEFs. These objective functions take a_g as the input.

$S_{aC}(t, T)$ denotes endurance time excitations target acceleration spectra, computed in the following equation:

$$S_{aC}\left(t,T\right) = g\left(t\right) \times S_a^{\text{target}}\left(T\right) \tag{2.13}$$

where $S_a^{\text{target}}(T)$ denotes ground motions' target acceleration spectra as the average acceleration spectra of ground motions. $g(t)$ is the *intensifying profile* which controls the shape of increasing acceleration spectra in time. In simulating existing endurance time excitations, a linear function had been adopted for the intensifying profile, where g should satisfy the following requirements:

1. g should be an ascending function.
2. $g(t_{\text{target}}) = 1$; t_{target} is time at which ETEFs match normalized GMs.
3. $g(0) = 0$, as initial condition.

$S_{uC}(t, T)$ is the endurance time excitations target linear displacement spectra, as computed by the following equation:

$$S_{uc}(t,T) = g(t) \times S_u^{\text{target}}(T) \tag{2.14}$$

where $S_u^{\text{target}}(T)$ denotes the normalized GMs target linear displacement spectra, which are the average displacement spectra of GMs.

CAV$_C(t)$ represents the endurance time excitations target CAV, and it is computed according to the following equation:

$$\text{CAV}_C(t) = h(t) \times \text{CAV}^{\text{target}} \tag{2.15}$$

where CAV^{target} stands for target cumulative absolute velocity associated with normalized ground motions, which is the average cumulative absolute velocity of ground motions; $h(t)$ is the increasing profile of cumulative absolute velocity in time and its requirements are similar to those mentioned for the function $g(t)$. It should be noted that the function $h(t)$ is not necessarily identical to the function $g(t)$ and should be determined.

α_{S_u} and α_{CAV} are weight factors which control the contribution of residuals, respectively, associated with displacement spectra and cumulative absolute velocity in the objective function. These factors are calculated by using the following equations:

$$\alpha_{S_u} = \frac{\displaystyle\int_0^{T_{\max}} \int_0^{t_{\max}} [S_{aC}(T,t)]^2 \, dt \, dT}{\displaystyle\int_0^{T_{\max}} \int_0^{t_{\max}} [S_{uC}(T,t)]^2 \, dt \, dT} \tag{2.16}$$

$$\alpha_{CAV} = \frac{\displaystyle\int_0^{T_{\max}} \int_0^{t_{\max}} [S_{aC}(T,t)]^2 \, dt \, dT}{\displaystyle\int_0^{T_{\max}} \int_0^{t_{\max}} [CAV_C(t)]^2 \, dt \, dT} \tag{2.17}$$

$S_a(t, T)$ is the acceleration spectra produced by endurance time excitations at time t and period T. Acceleration spectra of ETEFs are calculated by the following equation:

$$Sa(t,T) = \max\left(\left|\ddot{x}(\tau) + a_g(\tau)\right|\right) \quad 0 \leq \tau \leq t \tag{2.18}$$

where $\ddot{x}(\tau)$ is the relative acceleration response of a single degree of freedom with a period of T and a damping ratio of 5% under the endurance time excitations, and $a_g(\tau)$ is the acceleration time history of endurance time excitations.

$S_u(t, T)$ is the displacement spectra produced by endurance time excitations at time t and period T. The displacement spectra of endurance time excitations are calculated by the following equation:

$$S_u(t,T) = \max(|x(\tau)|) \quad 0 \leq \tau \leq t \tag{2.19}$$

where $x(\tau)$ is the relative displacement response of single degree of freedom with a period of T and damping ratio of 5% under the endurance time excitations.

CAV(t) is the cumulative absolute velocity produced at time t and by endurance time excitations as calculated by the following equation:

$$CAV(t) = \int_0^t |a_g(\tau)| d\tau \tag{2.20}$$

FIFTH GENERATION OF ENDURANCE TIME EXCITATIONS

In the fifth generation, acceleration spectra, nonlinear displacement spectra, and absorbed hysteretic energy are included in the generating process. These are the most advanced ETEFs as compared to the previous generations. It should be mentioned that absorbed hysteretic energy demand is used as an indicator of the cumulative damage related parameter. Linear increasing function for acceleration spectra is adopted, and the corresponding increasing function of nonlinear displacement spectra and hysteretic energy is mathematically computed.

The following equation is the objective function which calculates squared discrepancies between dynamic characteristics of endurance time excitations and target ground motions:

$$
\begin{aligned}
F_{ETEF}(a_g) = &\int_0^{T_{max}} \int_0^{t_{max}} \left\{ [S_a(t,T) - S_{aC}(t,T)]^2 \right\} dt\, dT \\
&+ \int_1^{\mu_{max}} \int_0^{T_{max}} \int_0^{t_{max}} \left\{ \begin{array}{l} \alpha_{u_m} \left[u_m(t,T,\mu) - u_{mC}(t,T,\mu) \right] \\ \alpha_{EH} \left[E_H(t,T,\mu) - E_{HC}(t,T,\mu) \right] \end{array} \right\} dt\, dT\, d\mu
\end{aligned} \tag{2.21}
$$

where $a_g(t)$ is the acceleration time history of ETEFs, $S_a(t, T)$, $u_m(t, T, \mu)$, and $E_H(t, T, \mu)$ are dynamic characteristics of endurance time excitations, whereas $S_{aC}(t, T)$, $u_{mC}(t, T, \mu)$, and $E_{HC}(t, T, \mu)$ are dynamic characteristics of ground motions. These parameters will be explained and discussed in the following lines.

$S_a(t, T)$ here stands for the acceleration spectra of endurance time excitation at time t and period T, which is calculated as follows:

$$S_a(t,T) = \max(|\ddot{x}(\tau) + a_g(\tau)|) \quad 0 \leq \tau \leq t \tag{2.22}$$

In this case, $\ddot{x}(\tau)$ is the relative acceleration response of single degree of freedom with a period of T and a damping ratio of 5% under ETEF and $a_g(\tau)$ is the acceleration time history of endurance time excitation.

Then, $S_{ac}(t,T)$ refers to target acceleration spectra of endurance time excitations at time t and period T, which is calculated using the following equation:

$$S_{ac}(t,T) = \frac{t}{t_{target}} \times S_a^{target}(T) \tag{2.23}$$

Here, $S_a^{target}(T)$ represents target acceleration spectra – which are the average of normalized ground motions and t_{target} is the time at which endurance time excitations should match normalized ground motions.

$u_m(t, T, \mu)$ is the displacement demand of nonlinear single degree of freedom with a period of T and yield strength of $F_y(T, \mu)$ subjected to endurance time excitations. Specifications of single degree of freedom are shown in Figure 2.1, where $F_y(T, \mu)$ is the minimum lateral strength capacity that a single degree-of-freedom system with a period of T requires to avoid the average ductility ratio demands larger than μ under these sets of ground motions. μ and T are the ductility demand and the period of single degree-of-freedom systems. Figure 2.2 illustrates $F_y(T, \mu)$ for three different μs as a function of T. $F_y(T, \mu)$s were then compared with yield strengths proposed by Miranda (1993). Correlation between $F_y(T, \mu)$ calculated by Miranda (1993), and those computed for FEMAp695 records is satisfactory. The following equation explores the calculation of $u_m(t, T, \mu)$:

$$u_m(t,T,\mu) = \max\left(\left|u(\tau)\right|\right) \ 0 \leq \tau \leq t \tag{2.24}$$

where $u(\tau)$ is the displacement response subjected to ETEF.

$u_{mc}(t, T, \mu)$ is the target nonlinear displacement which is calculated by the following equation:

$$u_{mc}(t,T,\mu) = h_u(t,T,\mu) \times u_m^{target}(T,\mu) \tag{2.25}$$

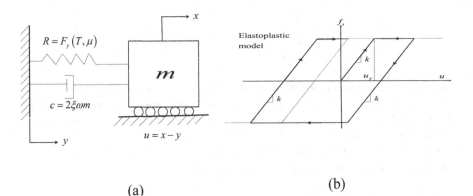

(a)   (b)

FIGURE 2.1   (a) Nonlinear single degree-of-freedom configuration and (b) behavior model.

As the equation reads, $u_m^{target}(T,\mu)$ is the nonlinear displacement imposed by normalized ground motions. $h_u(t, T, \mu)$ shows the variation of target in time. $h_u(t, T, \mu)$ is determined by multiplying ground motion scales a number of times and calculating demands for each scale factor (λ). λ can be transformed to time (t) by the transformation function of $\lambda = t/t_{target}$. Figure 2.3 depicts $h_u(t, T, \mu)$ for four combinations of T and μ.

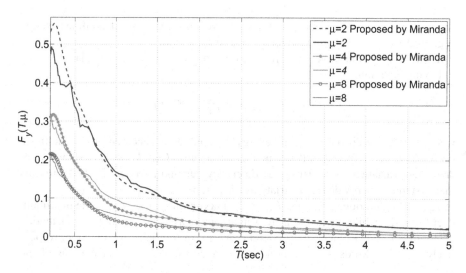

FIGURE 2.2 Yield strengths used in this study for generating ETEFs compared to those proposed by Miranda (1993).

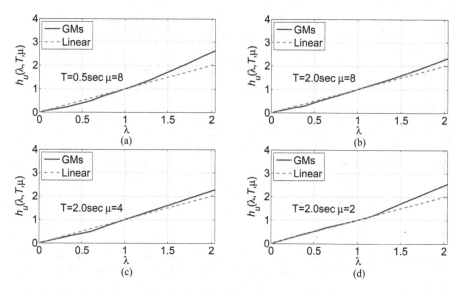

FIGURE 2.3 Time-dependent variations of target displacement demands: (a) $T=0.5$ second, $\mu = 8$; (b) $T=2.0$ seconds, $\mu = 8$; (c) $T=2.0$ seconds, $\mu = 4$; and (d) $T=2.0$ seconds, $\mu = 2$.

$E_h(t, T, \mu)$ is the hysteretic energy demand of nonlinear single degree of freedom under endurance time excitations and is calculated using the following equation:

$$E_H\left(t,T,\mu\right) = \int_0^t f_s\left(\tau\right)\dot{u}\left(\tau\right)d\tau - \frac{f_s\left(t\right)^2}{2k} \tag{2.26}$$

where f_s is the restoring force and $\dot{u}$ is the velocity response of nonlinear single degree-of-freedom system under endurance time excitation at t.

$E_{Hc}(t, T, \mu)$ is the target hysteretic energy demand which is calculated by the following equation:

$$E_{Hc}\left(t,T,\mu\right) = h_{EH}\left(t,T,\mu\right) \times E_H^{\text{target}}\left(T,\mu\right) \tag{2.27}$$

where $E_H^{\text{target}}(T, \mu)$ is the hysteretic energy demand under recorded ground motions. $h_{EH}(t, T, \mu)$ is determined in the same manner as $h_u(t, T, \mu)$. $h_{EH}(t, T, \mu)$ shows time-dependent variations of target hysteretic energy demand in time. Figure 2.4 reveals that hysteretic energy demand variations are parabolic.

$t_{\max}$ is the duration of ETEFs, $T_{\max}$ is the maximum period, $\mu_{\max}$ is the maximum ductility ratio, while α_{um} and α_{EH} are weight factors associated with the contribution of nonlinear displacement and hysteretic energy in the objective function, respectively. In other words, these coefficients balance the relative contributions of the

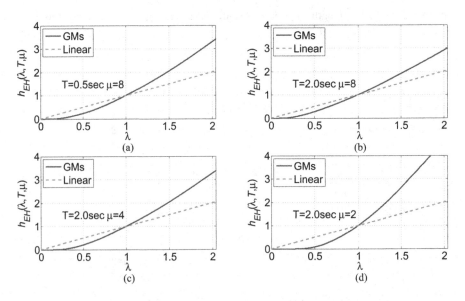

FIGURE 2.4 Time-dependent variation of target hysteretic energy demands: (a) $T=0.5$ second, $\mu = 8$; (b) $T=2.0$ seconds, $\mu = 8$; (c) $T=2.0$ seconds, $\mu = 4$; and (d) $T=2.0$ seconds, $\mu = 2$.

above-mentioned components. It should be noted that the metric units are employed. These factors are computed according to the following equations:

$$\alpha_{u_m}(\mu) = \frac{\displaystyle\int_0^{T_{max}}\int_0^{t_{max}} \left[S_{aC}(T,t)\right]^2 dt\, dT}{\displaystyle\int_0^{T_{max}}\int_0^{t_{max}} \left[u_{mc}(t,T,\mu)\right]^2 dt\, dT} \tag{2.28}$$

$$\alpha_{E_H}(\mu) = \frac{\displaystyle\int_0^{T_{max}}\int_0^{t_{max}} \left[S_{aC}(T,t)\right]^2 dt\, dT}{\displaystyle\int_0^{T_{max}}\int_0^{t_{max}} \left[EH_c(t,T,\mu)\right]^2 dt\, dT} \tag{2.29}$$

REFERENCES

Bolt, B. A. (1973). Duration of strong ground motion. In 5th *World Conference* on *Earthquake Engineering*. Rome (pp. 1304–1313).

Bommer, J. J., & Martinez-Periera, A. (1999). The effective duration of earthquake strong motion. *Journal of Earthquake Engineering*, 3(2), 127–172. https://doi.org/10.1080/13632469909350343

Campbell, K. W., & Bozorgnia, Y. (2012). Cumulative absolute velocity (CAV) and seismic intensity based on the PEER-NGA database. *Earthquake Spectra*, 28(2), 457–485. https://doi.org/10.1193/1.4000012

Electrical Power Research Institute. (1991). Standardization of Cumulative Absolute Velocity. EPRI Report RP3096-1. Electric Power Research Institute, Palo Alto, CA.

Hancock, J., & Bommer, J. J. (2006). A state-of-knowledge review of the influence of strong-motion duration on structural damage. *Earthquake Spectra*, 22(3), 827–845. https://doi.org/10.1193/1.2220576

Miranda, E. (1993). Site-dependent strength-reduction factors. *Journal of Structural Engineering*, 119(12), 3503–3519. https://doi.org/10.1061/(ASCE)0733-9445(1993)119:12(3503)

Wang, G., Wang, Y., Lu, W., Yan, P., Zhou, W., & Chen, M. (2016). A general definition of integrated strong motion duration and its effect on seismic demands of concrete gravity dams. *Engineering Structures*, 125, 481–493. https://doi.org/10.1016/j.engstruct.2016.07.033

3 Optimization Variable Spaces[1]

REVIEW

Various variable spaces can be used in the production process of ETEFs. In this chapter, some common variable spaces for endurance time excitation simulation are explained. The influence of optimization space on simulated excitations is also discussed. Finally, the results of simulated excitations are presented. This chapter focuses on variable definition types. It is worth mentioning that the solution method is independent of variable definition type. The adopted variable space can affect the efficiency of the ETEF generation and also the quality of produced ETEFs. It should be noted that the final ETEFs are presented in the form of time series regardless of the variable space used in their generation.

OPTIMIZATION VARIABLE SPACE FOR SIMULATING ENDURANCE TIME EXCITATIONS

As stated in Chapters 1 and 2, the optimization procedure is employed to simulate endurance time excitations. In the optimization context, equations are expressed in terms of objective function and variables are represented in terms of decision variables. In the literature, decision variables are also called optimization variables and design variables. Representation types of endurance time excitations characterize optimization variables. Three different endurance time excitation representations are presented as listed below:

- Time domain
- Wavelet representation
- Increasing Sine functions

TIME DOMAIN

One of the most common methods for endurance time excitations representation is to use acceleration data points. This simulation space is also referred to as time domain. Primitive endurance time excitations (series "a", "b", "c", "d", "h", "g", and "I") are

---

[1] Chapter sources:
 1. Mashayekhi, M., Estekanchi, H. E., & Vafai, H. (2019) Simulation of endurance time excitations using increasing sine functions. *International Journal of Optimization in Civil Engineering*, 9(1), 65–77.
 2. Mashayekhi, M., Estekanchi, H. E., & Vafai, H. (2018) Simulation of endurance time excitations via wavelet transform. *Iranian Journal of Science and Technology, Transactions of Civil Engineering.* https://doi.org/10.1007/s40996-018-0208-y

DOI: 10.1201/9781003216681-3

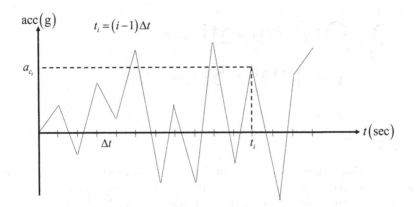

FIGURE 3.1 Definition of decision variables in time-domain optimization space.

built in this optimization space. Decision variables in this space are acceleration data points. Definition of variables in time-domain space is illustrated in Figure 3.1. Equations of endurance time excitation simulation in time-domain space are brought in Equation (3.1). As expressed in the beginning of this chapter, the solution method is independent of variable definition type.

$$\text{Find } \{a_{c_i}\}$$
$$\text{To Minimize } F_{ETEF}\{x\} \tag{3.1}$$

where a_{c_i} is the acceleration data point of endurance time excitations and $F_{ETEF}\{x\}$ is the objective function that is thoroughly explained in Chapter 2. It should be mentioned again that the solution method is independent of variable definition.

WAVELET DECOMPOSITION

BASIC THEORY OF WAVELET ANALYSIS

Unlike the Fourier analysis, which uses sinusoidal waves as tools for decompositions, the wavelet transform utilizes scaled and translated versions of scaling function $\phi(t)$ and wavelet function $\psi(t)$. Different scaled versions cover different frequency resolutions, and different translated versions represent different time positions. Basis functions and scaling functions of the wavelet transform are given in the following equations:

$$\phi_{j,k}(t) = 2^{j/2}\phi\left(2^j t - k\right) \tag{3.2}$$

$$\psi_{j,k}(t) = 2^{j/2}\psi\left(2^j t - k\right) \tag{3.3}$$

where j is the dilatation parameter and k is the position parameter. ϕ refers to the scaling function, while ψ denotes the wavelet function or the mother wavelet.

As represented by Equation (3.4), scaling function (ϕ) is derived from the dilation equation:

$$\phi(t) = \sum_{k=0}^{N-1} c_k \phi(2t - k) \tag{3.4}$$

where c_k refers to wavelet coefficients. The corresponding wavelet function is derived from the following equation:

$$\psi(t) = \sum_k (-1)^k c_k \phi(2t + k - N + 1) \tag{3.5}$$

Coefficients from contributing wavelets (after wrapping $f(t)$ signals) can be computed by Equations (3.6) and (3.7) (Newland, 1993). The $f(t)$ signal can be reconstructed by Equation (3.8):

$$a_{2^j+k} = 2^j \int_0^l f(t)\psi\left(2^j t - k\right) dt \tag{3.6}$$

$$a_0 = \int_0^l f(t)\phi(t) dt \tag{3.7}$$

$$f(t) = a_0\phi(t) + a_1\psi(t) + \begin{bmatrix} a_2 & a_3 \end{bmatrix} \begin{bmatrix} \psi(2t) \\ \psi(2t-1) \end{bmatrix} +$$

$$\begin{bmatrix} a_4 & a_5 & a_6 & a_7 \end{bmatrix} \begin{bmatrix} \psi(4t) \\ \psi(4t-1) \\ \psi(4t-2) \\ \psi(4t-3) \end{bmatrix} \tag{3.8}$$

$$... + a_{2^j+k}\psi\left(2^j t - k\right) + ...$$

When necessitated by circumstances, all translated and scaled wavelets and scaling functions shall be wrapped around the signal length in Equations (3.6)–(3.8). It is demonstrated that after wrapping, the scaling function becomes constant. Discrete wavelet transform (DWT) is an algorithm used to compute the integrations mentioned above and is also applicable to signals sampled at equal intervals. Mallat (1989) has developed an algorithm to derive these coefficients without directly solving the convolution of Equations (3.6) and (3.7).

Scaling function $\phi(t)$ and wavelet function $\psi(t)$ of db12 (Daubechies, 1992) which are computed by Equations (3.8) and (3.9) are shown in Figure 3.2.

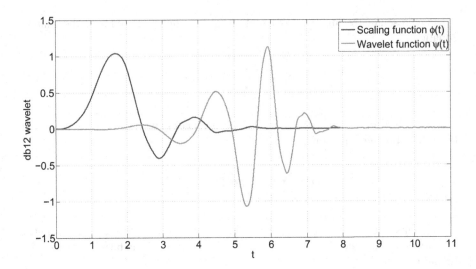

FIGURE 3.2   Scaling function $\phi(t)$ and mother wavelet function $\psi(t)$ of db12.

SIMULATION METHOD

Simulating in Wavelet Transform Space

Endurance time excitations in DWT space are represented by Equation (3.9). In this case, $a_{j,k}$ must be specified by the optimization process for simulation procedure. In Equation (3.9), a signal consisting of 2^M data points, where M is an integer, is considered. DWT requires 2^M wavelet coefficients to fully describe this signal. DWT decomposes a signal into $M+1$ levels, where levels are denoted as i and numbered as $i = -1, 0, 1, \ldots, M-1$ (Newland, 1993). In this equation, signals over time duration $t_{\max}$ are sampled at N equally spaced time-sequenced Δt; N is power of 2:

$$a_g = \sum_{j=0}^{M-1} \sum_{k=1}^{2^j} a_{j,k} \psi_{j,k}(t) \tag{3.9}$$

Since signals are represented by corresponding DWT coefficients, the objective function must be defined as a function of DWT coefficients $(a_{j,\,k})$. In this regard, SDOF responses under signals must be expressed as functions of DWT coefficients. The following equation takes wavelet coefficients of a signal and returns responses of the SDOF system with a fundamental period of T:

$$\ddot{x}(t) = \sum_{j=0}^{M-1} \sum_{k=1}^{2^j} a_{j,k} \ddot{x}_{j,k}(t,T) \tag{3.10}$$

where $\ddot{x}_{j,\,k}(t, T)$ is the acceleration response of the SDOF system with a period of T under a signal at which all wavelet coefficients are zero except $a_{j,k}$.

In order to generate endurance time excitations in the DWT space, the procedure depicted in Figure 3.3 is followed. Here, signals are represented by Equation (3.9), and SDOF responses used in evaluation of the objective function are computed by Equation (3.9).

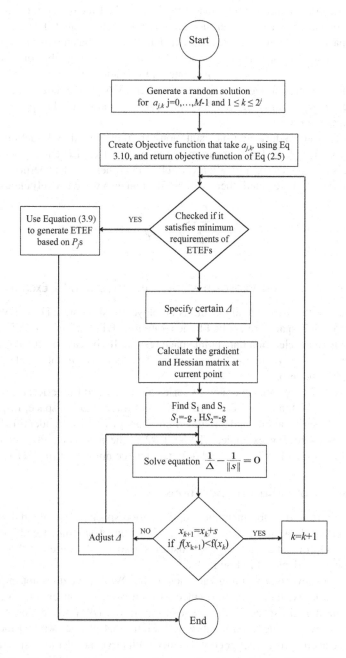

FIGURE 3.3 Algorithm of generating endurance time excitations in the DWT space.

Simulating in Filtered Transformed Space

Reducing optimization of variable dimensions is one of the major benefits of producing endurance time excitations in the DWT space. Reduction of variables in this method leads to considerably less computational efforts in generating process. While generating endurance time load functions in the time domain, the number of optimization variables and acceleration data points must be equal. Thus, in the time-domain space, reduction of optimization variables cannot be employed.

The reduction of variables is carried out by limiting the number of wavelet levels representing signals. For example, one can consider only the first 512 wavelet coefficients to represent signals and set the other 1536 wavelet coefficients to zero. Therefore, $[a_1, a_2, ..., a_{512}]$ are the decision variables and should be specified during the optimization process.

In DWT space, endurance time excitations are represented by Equation (3.11). In this equation, M_{cut} is the filtering parameter where DWT coefficients associated with lower levels are taken as optimization variables, and other wavelet coefficients are set to zero. If $M_{cut} = 8$ is selected, there will be 512 nonzero wavelet coefficients:

$$a_g = \sum_{j=0}^{M_{cut}} \sum_{k=1}^{2^j} a_{j,k} \psi_{j,k}(t) \tag{3.11}$$

THE SIGNIFICANCE OF THE PROPOSED METHOD IN SIMULATING ET EXCITATIONS

In this section, the affronted algorithm is employed to simulate ETEFs. DWT space and filtered DWT space are applied to generate new ETEFs. These ETEFs are then compared with acceleration data point space results. In the case of the filtered DWT space, the first 512 wavelet coefficients are used as decision variables and other coefficients are set to zero.

Three ETEF series are generated: A series, D series, and FD series, the optimization space of which are time-domain, DWT, and filtered DWT space, respectively. The duration, series name, and simulated numbers of ETEFs are identified in standard ETEFs name. For example, ETA20FD02 is the name of a 20-second ETEF simulated in filtered DWT space and (02) denotes the number of the ETEF.

CONFORMANCE OF GENERATED EXCITATIONS

In this subsection, the aforementioned optimization spaces are compared with each other in terms of the normalized objective function of simulated endurance time excitations. Optimization convergence histories associated with these three optimization spaces are depicted in Figure 3.4.

Figure 3.4 shows that defining variables in the DWT space does not improve the endurance time excitations conformance as compared to conventional generating spaces (time-domain space). It can be seen that in the DWT space, convergence is not obtained even by iteration number 300. Figure 3.4 also shows that the filtered DWT space accomplishes better conformance with target acceleration spectra. It can be seen that convergence is obtained in the iteration number of 130.

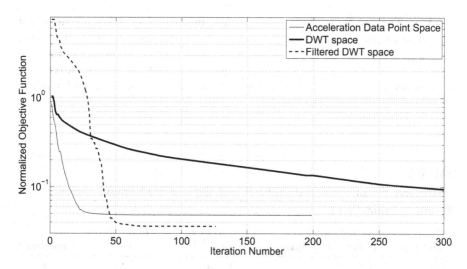

FIGURE 3.4 Optimization convergence history for the time-domain, DWT, and filtered DWT spaces.

SENSITIVITY OF GENERATED EXCITATIONS TO INITIAL OPTIMIZATION POINTS

Another important issue regarding optimization space is the sensitivity of the simulated endurance time excitations' conformance with target acceleration spectra to initial points. Minimum sensitivity is desired in providing a reliable simulating method. Sensitivity is quantified by dispersion of normalized objective function values associated with simulated excitations generated by different random initial points. The dispersion is computed according to the following equation:

$$s = \sqrt{\frac{1}{n_E - 1} \sum_{i=1}^{n_E} \left(F_{N,i} - \bar{F}_N \right)^2} \tag{3.12}$$

where n_E is the number of random initial points used in order to explore the sensitivity to initial points, $F_{N,i}$ is the normalized objective function value of simulated excitations based on the ith initial point, and $\overline{F_N}$ represents the average simulated ETEFs.

The results are summarized in Table 3.1. According to Table 3.1, the optimization in the DWT space increases dependence on initial points, while dependence is decreased in the filtered DWT space in comparison with the time-domain optimization. It can be observed that the time elapse of the optimization process in the filtered DWT space is about 20% of that associated with the time-domain optimization. On the contrary, the DWT space increases time elapse of the optimization process. The decrease in computational demand of filtered DWT space is due to decrease in the size of optimization space and decrease in the required iteration numbers. The number of function evaluations in trust region reflective method at each iteration is the size of optimization space owing to numerical calculation of objective function gradient. Therefore, the total number of objective function evaluations of each optimization space is equal to the multiplication of the required iteration numbers and the size of optimization space. Both the mentioned numbers are decreased in filtered DWT space.

TABLE 3.1

Characteristics of Simulated ETEFs, Which Explore the Dependence of Results to Initial Points

ETETs Series	Number	Opt. Space	Normalized Obj. Function	Time Elapse (hour)	Dispersion
ET20A	01	Time	0.0483	10.4	0.0486
	02	domain	0.1360	9.4	
	03		0.0558	10.6	
ET20D	01	DWT	0.8806	16.2	0.4334
	02		0.0796	17.4	
	03		0.1933	15.5	
ET20FD	01	Filtered	0.0364	1.9	0.002
	02	DWT	0.0400	1.5	
	03		0.0398	1.7	

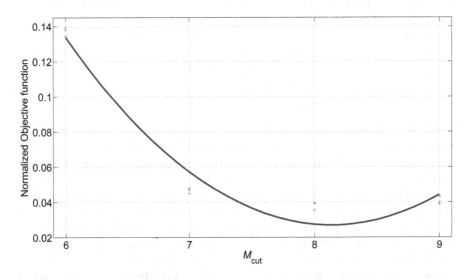

FIGURE 3.5   Normalized objective functions vs. filtering levels in the DWT space.

Optimum Level of Filtering

The optimal generating scenario must be determined. In other words, the filtering parameter (M_{cut}) must be determined.

It should be noted that when no filtering is considered, several insignificant variables cause numerical problems in the generating process. However, if excessive filtering is performed, important information of signals is discarded and the global minimum point cannot be achieved.

It should be considered that when $M = 10$ is used, it implies that no filtering has been conducted, and all wavelet coefficients are incorporated in the optimization process. Moreover, when $M_{cut} = 5$ is used, convergence is not achieved, and therefore, this level of filtering cannot be considered. So, in order to find the optimal value of M_{cut}, four filtering thresholds are investigated, $M_{cut} = 6$, 7, 8, and 9. For each threshold, the optimization algorithm is executed three times using three random initial points. The normalized objective functions are plotted vs. filtering levels as shown in Figure 3.5. It can be seen that $M_{cut} = 8$ is the optimum filtering level.

RESULTS

The acceleration time history of ETA20FD01~03 is shown in Figure 3.6. Figure 3.7 compares the ETA20FD01~03 acceleration spectra with the target spectra at different times, i.e., 5, 10, 15, and 20 seconds. In addition, Figure 3.8 compares the ET20FD01~03 maximum absolute acceleration responses with the targets at different periods, i.e., $T = 0.03$ second, $T = 0.8$ second, $T = 2.0$ seconds, and $T = 3.0$ seconds. Results show striking consistency between the endurance time excitation spectra and the target spectra.

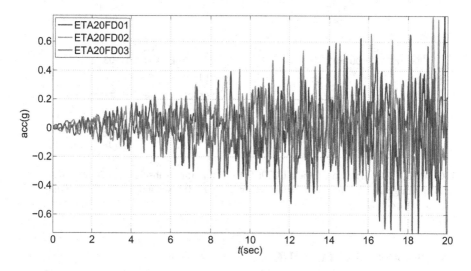

FIGURE 3.6   ETA20FD01~03 acceleration time histories.

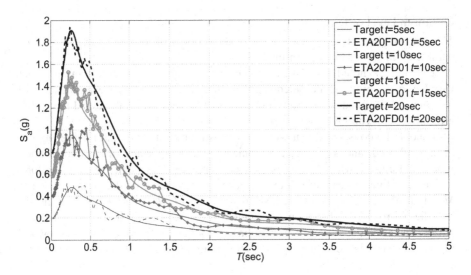

FIGURE 3.7 Comparison of the ETA20FD01~03 acceleration spectra and the targets at *t* = 5, 10, 15, and 20 seconds.

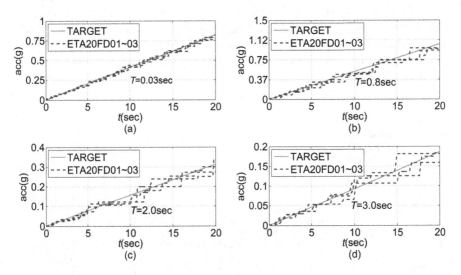

FIGURE 3.8 Comparison of the ETA20FD01~03 maximum absolute acceleration responses with the targets at periods (a) $T = 0.03$ second, (b) $T = 0.8$ second, (c) $T = 2.0$ seconds, and (d) $T = 3.0$ seconds.

INCREASING SINE FUNCTIONS

In this section, endurance time excitations are represented by using increasing sine functions. In fact, increasing sine functions are used as basis function. The schematic

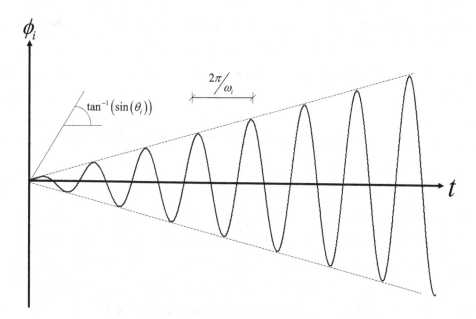

FIGURE 3.9 Increasing sine basis function.

of this basis function is shown in Figure 3.9. This signal representation is presented in the following equation:

$$a_g(t) = \sum_{i=1}^{n_f} a_i \phi_i(t) = \sum_{i=1}^{n_f} a_i t \sin(\omega_i t + \theta_i) \tag{3.13}$$

where $\phi_i(t)$ is the ith basis function which is an increasing sine function shown in Figure 3.9, a_i is the contributing factor of the ith basis function in acceleration time history of endurance time excitations that must be determined during optimization process, ω_i and θ_i are the angular frequency and the phase angles of the ith increasing sine functions, respectively, and n_f is the number of considered increasing sine functions.

The frequency of the ith increasing sine function is calculated according to Equation (3.14). In this equation, frequencies of considered increasing sine functions are distributed logarithmically between the maximum and the minimum considered frequencies:

$$\omega_i = 10^{\frac{\log(\omega_{min}) \times (n_f - i) + \log(\omega_{max}) \times (i-1)}{n_f - 1}} \tag{3.14}$$

where ω_{min} and ω_{max} denote the minimum and the maximum considered frequencies for increasing sine functions.

Simulating endurance time excitations is to find values of these coefficients so as to minimize the objective function of the problem. Objective function value is also denoted as cost function value in this paper. The problem of simulating endurance time excitations is summarized as below:

$$\text{Find } \{x\}_{1\times 2n_f} = \left[a_1, a_2, ..., a_{n_f}, \phi_1, \phi_2, ..., \phi_{n_f} \right]$$

$$\text{To Minimize } F_{ETEF} \{x\}$$

The algorithm for implementing the new optimization space is depicted in Figure 3.10.

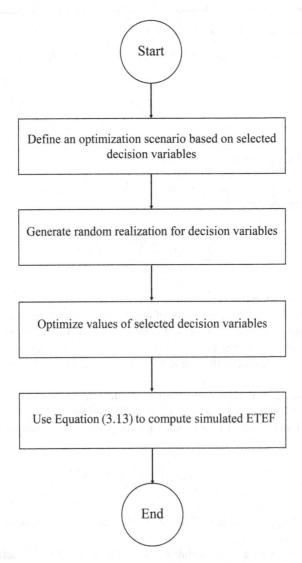

FIGURE 3.10 The proposed algorithm for implementation of new basis function in simulating ETEFs.

PARAMETER TUNING

In this section, 40-second endurance time excitations are simulated by using the proposed method. One advantage of the proposed method over conventional endurance time excitations simulating procedure is the possibility to adjust the number of optimization variables. In this study, eight optimization scenarios are defined. The number of optimization variables is proportional to the number of considered increasing sine functions. If phase angle is not considered in the process, the number of optimization variables will be equal to the number of considered increasing sine functions. Otherwise, the number of optimization variables will be equal to twice the number of considered increasing sine functions. In this study, 1.3 and 314.2 rad/sec are assigned to ω_{min} and ω_{max}, respectively. The considered scenarios are presented in Table 3.2.

For each scenario, the optimization algorithm depicted in Figure 3.10 is executed by using three different initial random motions. The used initial motions are generated randomly and the associated probability distribution parameters are presented in Table 3.3. Results of these runs are summarized in Table 3.4. Results show that ETEF-IS-04 brings about more accurate endurance time excitations. In addition, it can be observed that including phase angles as optimization variables could not improve the accuracy of simulated ETEFs and leads to the increase the cost function values.

TABLE 3.2
Optimization Scenarios for Simulating ETEFs Based on the Proposed Method

Scenario	n_f	Phase Angle Inclusion	Optimization Variable Number
ETEF-IS-01	300	No	300
ETEF-IS-02	600	No	600
ETEF-IS-03	900	No	900
ETEF-IS-04	1200	No	1200
ETEF-IS-05	300	Yes	600
ETEF-IS-06	600	Yes	1200
ETEF-IS-07	900	Yes	1800
ETEF-IS-08	1200	Yes	2400

TABLE 3.3
The Distribution Functions of Variables for Generating Initial Random Motions

Variables	Distribution Type	Distribution Parameters	
		Parameter	Value
Amplitudes	Normal	Median	0
		Standard deviation	0.005
Phase angles	Uniform	Lower bound	0
		Upper bound	6.28

TABLE 3.4
Simulated ETEFs' Results of Defined Scenarios

Scenario	Run Number	Cost Function		Computational Time (seconds)	
		Values	Average	Values	Average
ETEF-IS-01	01	347.2	341.2	6710	8522
	02	360.1		6239	
	03	316.4		12,618	
ETEF-IS-02	01	265.4	251.3	15,294	14,185
	02	265.3		15,345	
	03	223.1		11,917	
ETEF-IS-03	01	212.3	216.0	17,250	19,915
	02	221.4		21,926	
	03	214.3		20,569	
ETEF-IS-04	01	173.8	198.8	15,802	16,473
	02	214.1		14,945	
	03	208.6		18,672	
ETEF-IS-05	01	451.8	396.8	58,280	58,145
	02	379.2		58,146	
	03	359.3		58,010	
ETEF-IS-06	01	390.1	369.2	102,538	120,741
	02	332.5		102,842	
	03	385.1		156,843	
ETEF-IS-07	01	336.6	410.1	167,256	166,666
	02	426.4		166,216	
	03	467.3		166,528	
ETEF-IS-08	01	373.3	364.4	462,412	464,456
	02	367.4		465,357	
	03	352.4		465,599	

In order to investigate the efficiency of the proposed method, three endurance time excitations are simulated in time domain. It should be considered that simulating excitations is currently performed in time domain. Further investigation is conducted by comparing the proposed method results with simulating endurance time excitations in wavelet transform space optimization. Endurance time excitations simulated in time domain are hereafter denoted by ETEF-T, and ETEFs simulated in discrete wavelet transform space are hereafter denoted by ETEF-W. Two scenarios are defined in discrete wavelet transform space, namely ETEF-W-01 and ETEF-W-02. In ETEF-W-01, the first 512 wavelet coefficients are considered as optimization variables, while in ETEF-W-02 the first 1024 wavelet coefficients are considered as optimization variables. Simulated excitations' results are summarized in Table 3.5. Results show that the scenario ETEF-W-02 creates more accurate excitations among the considered conventional scenarios.

TABLE 3.5

Results of ETEFs Simulated in Time Domain and Discrete Wavelet Transform Space

Scenario	Run Number	Cost Function		Computational Time	
		Values	Average	Values	Average
ETEF-T	01	5964.4	4448.3	242,367	242,654
	02	4704.8		244,227	
	03	2675.6		241,369	
ETEF-W-01	01	366.2	320.4	31,644	25,901
	02	309.3		21,777	
	03	285.6		24,282	
ETEF-W-02'	01	232.6	235.3	56,697	56,302
	02	248.6		59,325	
	03	224.7		52,885	

TABLE 3.6

The Proposed Optimization Space vs. the Conventional Approach

Parameter	Conventional Approach	Proposed Method
Average cost function	235.3	198.8
Standard deviation cost function	12.2	21.8
Computational time (seconds)	56,302	16,473
Best cost	224.7	173.8
Worst cost	248.6	214.1

Table 3.6 compares the proposed method results with the conventional approach ones. Results show that the proposed method improves the accuracy of simulated excitations by 15.5%. The proposed method also decreases the required computational time by 71%. However, the standard deviation of cost functions associated with the proposed method increases by 44%.

In order to quantify the accuracy of the simulated excitations, normalized residuals are defined as in Equations (3.15)–(3.17). NRR integrates residuals at all times and periods. The residuals at each time are integrated at all periods and then are normalized. This normalization method avoids residuals domination where response spectra values are little and division by little numbers occurs. Normalized residuals express the accuracy of endurance time excitations in

TABLE 3.7

Normalized Residuals of ETEFs Simulated by the Proposed Method vs. Conventional Approaches

Parameter	The Proposed Method	Conventional Approach
NRR_{Sa}	6.4%	7.8%
NRR_{Sd}	22.5%	48%
NRR_{Sv}	13.4%	32.8%

percent. Therefore, it is an acceptable measure to investigate the efficiency of endurance time excitations.

$$NRR_{S_a} = \frac{1}{t_{max}} \int_0^{t_{max}} \left(\frac{\int_{T_{min}}^{T_{max}} \left| \left(S_a\left(T,t\right) - S_{ac}\left(T,t\right) \right) \right| dT}{\int_{T_{min}}^{T_{max}} S_{ac}\left(T,t\right) dT} \right) dt \qquad (3.15)$$

$$NRR_{S_d} = \frac{1}{t_{max}} \int_0^{t_{max}} \left(\frac{\int_{T_{min}}^{T_{max}} \left| \left(S_d\left(T,t\right) - S_{dc}\left(T,t\right) \right) \right| dT}{\int_{T_{min}}^{T_{max}} S_{dc}\left(T,t\right) dT} \right) dt \qquad (3.16)$$

$$NRR_{S_v} = \frac{1}{t_{max}} \int_0^{t_{max}} \left(\frac{\int_{T_{min}}^{T_{max}} \left| \left(S_v\left(T,t\right) - S_{vc}\left(T,t\right) \right) \right| dT}{\int_{T_{min}}^{T_{max}} S_{vc}\left(T,t\right) dT} \right) dt \qquad (3.17)$$

where NRR_{S_a}, NRR_{S_d}, and NRR_{S_v} are normalized residuals associated with acceleration spectra, displacement spectra, and velocity spectra, respectively.

Table 3.7 compares normalized residuals of endurance time excitations simulated by the proposed method and excitations simulated by conventional approaches. Results show somewhat improvement in excitations simulated by the proposed method.

ACCURACY OF EXCITATIONS SIMULATED WITH THE PROPOSED METHOD

In this section, endurance time excitations simulated by the ETEF-IS-04 are presented and are explored. These excitations are denoted by ETEF-IS. Convergence history of simulating ETEF-IS is depicted in Figure 3.11. It should be mentioned that the scale of ordinate and abscissa of this figure is logarithmic. Acceleration time history of ETEF-IS is shown in Figure 3.12. Acceleration spectra of ETEF-IS are compared with targets in Figure 3.13 at four times, $t = 15, 25, 35,$ and 40 seconds. This figure shows the acceptable correspondence between ETEF-IS acceleration spectra and targets. This fact proves the efficiency of the proposed method.

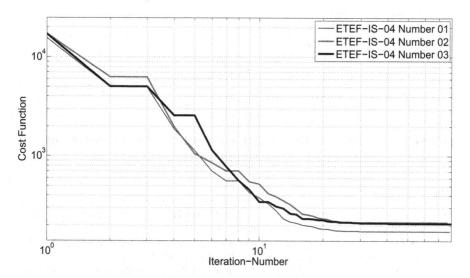

FIGURE 3.11 Convergence history of simulating ETEF-IS.

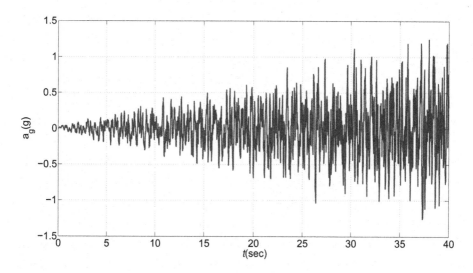

FIGURE 3.12 Acceleration time history of ETEF-IS.

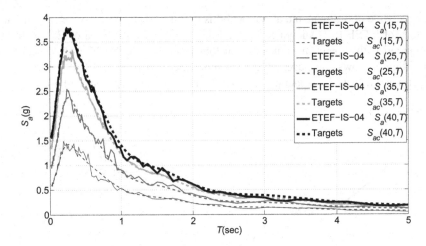

FIGURE 3.13 Comparison of acceleration spectra of ETEF-IS with targets at time $t = 15$ seconds, $t = 25$ seconds, $t = 35$ seconds, and $t = 40$ seconds.

REFERENCES

Daubechies, I. (1992). *Ten Lectures on Wavelets*. CBMS-NSF Conference Series in Applied Mathematics, vol 61. Montpelier, VT: SIAM.

Mallat, S. G. (1989). Multiresolution approximations and wavelet orthonormal bases of $L^2(R)$. *Transaction of American Mathematical Society*, *315*, 67–87.

Newland, D. (1993). *Random Vibrations, Spectral and Wavelet Analysis* (3rd Edition). New York: Longman Scientific & Technical.

4 Generating ETEFs Based on Linear Spectra[1]

REVIEW

In this chapter, the procedure of simulation endurance time excitations based on linear analysis is explained. Simulation procedure of series "a" and "lc" is explained. "a" is second- and "lc" is fourth-generation type. Duration consistency is included in simulating fourth-generation endurance time excitations. Cumulative absolute velocity as a duration-related parameter is considered in the simulation of "lc" series. The objective functions of "a" series and "lc" series are provided in Chapter 2. "a" series is generated in time-domain space, while "lc" series is generated in filtered discrete wavelet transform space. While ETA20lc series belonging to the fourth generation is considered a relatively advanced ETEF series, linear spectral matching in long period could still be used to achieve good results in nonlinear response range. These spaces are thoroughly described in Chapter 3. Detailed explanation of solution method (i.e., optimization setup) is provided in this chapter.

SIMULATION OF "A" SERIES

A typical code compliant accelerogram that corresponds to Standard No. 2800 of the Iranian National Building Code (INBC) has been used to define the target response spectra, as follows:

$$\begin{cases} B = 1 + 1.05\left(\dfrac{T}{0.1}\right) & T < 0.1\,\text{second} \\[2mm] B = 2.5 & 0.1\,\text{second} \leq T \leq 0.5\,\text{second} \\[2mm] B = 2.5\left(\dfrac{0.5}{T}\right)^{\frac{2}{3}} & 0.5\,\text{second} \leq T \end{cases} \tag{4.1}$$

$$S_{aC} = \frac{0.35BI}{R}$$

where I is the importance factor considered to be 1.0 and R is the response reduction factor that has not been applied (assumed to be equal to 1.0). Unconstrained nonlinear numerical optimization was used to solve the problem considering 2048 (2^{11})

[1] Chapter source: Mashayekhi, M., Estekanchi, H. E., Vafai, A., & Mirfarhadi, S. A. (2018) Simulation of cumulative absolute velocity consistent endurance time excitations. *Journal of Earthquake Engineering*, 25(5), 892–917.

DOI: 10.1201/9781003216681-4

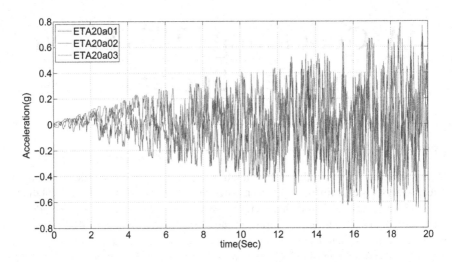

FIGURE 4.1   Series "a" excitation functions.

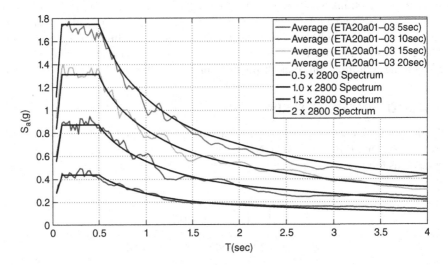

FIGURE 4.2   Comparison between acceleration spectra of "a" series with target.

accelerogram data points as optimization variables. Second-generation endurance time excitations were used to define the initial values of the variables in the optimization process. 200 different periods which are logarithmically distributed were assumed in the numerical optimization. A typical accelerogram produced by this procedure is depicted in Figure 4.1. Comparison between acceleration spectra of "a" series with targets is provided in Figure 4.2. Compatibility of second-generation excitations with targets is considerably improved as compared to the first generation. Convergence history of simulating a second-generation excitation is shown in Figure 4.3. This figure indicates that it is not possible to reach a better result with further trial and error and the optimization problem converged.

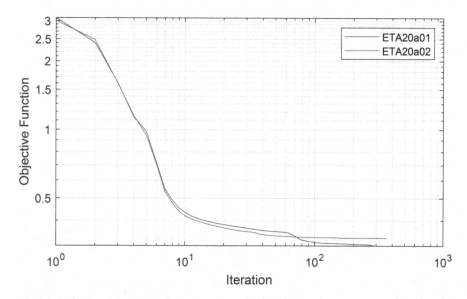

FIGURE 4.3 Convergence history of optimization problem.

SIMULATION OF "LC" SERIES

PROBLEM DEFINITION

In the simulation of "lc" series, CAV consistency is included. CAV consistency is referred to the condition in which the cumulative absolute velocity conditioned to a given spectral acceleration at the first mode period of structures is consistent between the endurance time method and the incremental dynamic analysis. Variation function of CAVs vs. acceleration at structural first mode period is investigated for an existing endurance time excitation and a ground motion which is successively scaled to cover a range of interest. This multiple scaling of a suite of ground motions is the common approach of incremental dynamic analysis. Variations of CAV vs. S_a ($T=1$ second) are illustrated for both the IDA and the ET method in Figure 4.4. It can be seen that this relationship is linear for IDA, while it appears parabolic in endurance time method. It can be observed that the ET method, by using current endurance tune excitations and the IDA, is inconsistent regarding the variation of CAV against acceleration at the first structural mode period. This inconsistency can be resolved by adjusting $g(t)$.

As explained before, $g(t)$ controls the increasing shape of acceleration spectra, while $h(t)$ controls the increasing shape of CAV in endurance time excitations. $g(t)$ must be determined in order to create a linear connection between CAV and $S_a(T)$, as stated in the following equation:

$$CAV = a * S_a\left(T_1, 5\%\right) \tag{4.2}$$

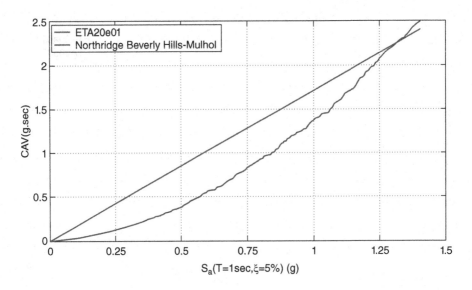

FIGURE 4.4 Variation comparison of CAV vs. S_a(T1) for Northridge Beverly Hills-Mulhol (one recorded GM) and ETA20e01 (one existing ETEF).

CAV can be approximated by $g(\tau)$, and therefore, Equation (4.2) changes to a differential equation, namely the following equation:

$$\int_0^t g(\tau)d\tau = a * g(t) \tag{4.3}$$

where a is a constant parameter.

The result of Equation (4.3), while satisfying initial conditions, is an exponential function. This exponential function is multiplied by the tangent hyperbolic function in order to satisfy the initial condition at zero ($g(0)=0$), as stated in Equation (4.4). It should be noted that the impact of tangent hyperbolic function in intensifying profile decreases in time. This matter can be attributed to the fact that the tangent hyperbolic comes to its asymptotes in about 1 second.

$$g(t) = b \tanh(\gamma t)e^{\alpha t} \tag{4.4}$$

where values of α, β, and γ are the profile intensifying rate parameters. The parameters used in this study are listed in Table 4.1.

With the intensifying profile of Equation (4.4), the function $h(t)$ is similar to the function $g(t)$. It was previously stated that these functions are not necessarily similar. For example, if linear function is adopted for the function $g(t)$, the function $g(t)$ will be parabolic and differs from the function $h(t)$.

Far-field ground motions proposed by FEMAP695 (2009) have been used in this study. These ground motions are recorded from large magnitude events ($M > 6.5$) at sites located greater than or equal to 10 km from fault rupture. This set includes records from soft rock, stiff sites, and shallow crustal sources. Individual records are

TABLE 4.1
Intensifying Profile Parameters

| ETEFs set | Duration Parameters | | Profile Parameters | | |
	t_{max} (seconds)	t_{target} (seconds)	β	γ	α
ETA20lc	20.48	10	0.5	0.25	0.0693
ETA40lc	40.96	20	0.25	0.25	0.0693

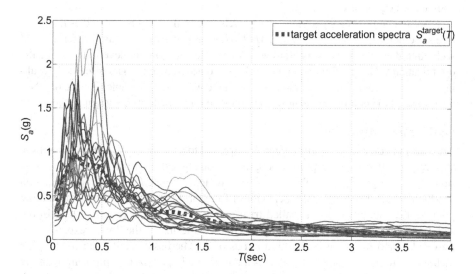

FIGURE 4.5 Ground motions target acceleration spectra.

normalized by their respective peak ground velocities. The procedure described in FEMAP695 (2009) is employed in this study. Normalization by peak ground velocity is a simple way to remove unwanted variability between records due to inherent difference in event magnitude, source type, and site condition while maintaining inherent aleatory variability for predicting seismic response assessment. $Sa_{target}(T)$ associated with these ground motions is depicted in Figure 4.5.

PROBLEM SOLUTION

Discretization

The objective functions provided in Chapter 2 for the fourth generation must be discretized to be solved, and it should be noted that the type of discretization manifests an impact in the results. Suppose that times are sampled at n points (t_j, $j = 1:n$) and periods are sampled at m points (T_i, $i = 1:m$). In this study, periods (T) are discretized at 130 points; 120 points in the interval [0.02, 5 seconds] which are distributed logarithmically; and 10 points in the interval [5, 50 seconds]. The time variable, t, is discretized at 4096 points in the time step of 0.01 seconds for 40-second endurance time excitations ($t_{max} = 40.95$ seconds) and at 2048 points in the time step 0.01 second for 20-second endurance time excitations ($t_{max} = 20.47$ seconds).

Signal Representation

Endurance time excitations are represented by their wavelet coefficients. Therefore, in order to generate new endurance time excitations, their wavelet coefficients must be specified. In case of a signal consisting of 2^M data points, where M is an integer, DWT requires 2^M wavelet coefficients to fully describe the signal. Mallat (1989) developed an algorithm to derive these 2^M coefficients directly needless of solving convolution equations.

DWT decomposes a signal into $M+1$ levels, where the level is denoted as i, and the levels are numbered $i = -1, 0, 1, \ldots, M-1$ (Newland, 1993). For example, for a signal with 2048 data points, there is 2048 wavelet coefficients with ten levels. To reduce the optimization variables, the first eight levels are considered and others discarded, resulting in a simplification of the problem. In fact, there will be 512 variables instead of 2048, meaning that the computations now carry on with one-fourth of the original load.

Optimization Algorithm

Nonlinear unconstrained optimization is employed to determine $a_{j,k}$. In this generation, trust-region reflective is employed as optimization algorithm. It is a simple yet powerful concept in optimization. The basic idea is to approximate f with a simpler function q, which reasonably reflects the behavior of function f in neighborhood N around the point x. This neighborhood is called trust region (Moré and Sorensen, 1983). In the standard trust region, q is defined by the first two terms of the Taylor expansion of f around x, and N is spherical with a radius of Δ. The basic equation of trust-region method is put in Equation (4.5). This is a constrained optimization problem. g and H are gradient and Hessian matrix of f at x, and s is the step size to be determined at each iteration. Algorithm of generating ETEFs is depicted in Figure 4.6.

$$\text{minimize} \left\{ \frac{1}{2} s^T H s + s^T g \right\} \text{ such that } \|s\| \leq \Delta \tag{4.5}$$

SIMULATED EXCITATIONS

20-Second Excitations

In this section, the proposed CAV-consistent generating method is applied in simulating 20-second endurance time excitations, as represented in Figure 4.7a. Figure 4.7b demonstrates the CAV vs. target, and as it can be seen, CAV consistency is not acceptably achieved. Such incompatibility can be attributed to the multiplication of the tangent hyperbolic function in the intensifying profile. The tangent hyperbolic function makes it possible to satisfy the initial condition of $g(t)$, though it affects CAV consistency. However, this negative impact of tangent hyperbolic function on CAV consistency decreases in time. Hence, it can be expected that CAV compatibility increases with time as the impact of tangent hyperbolic function fades away. This fact can be seen in Figure 4.7b where CAV consistency undergoes considerable improvement after 15 seconds.

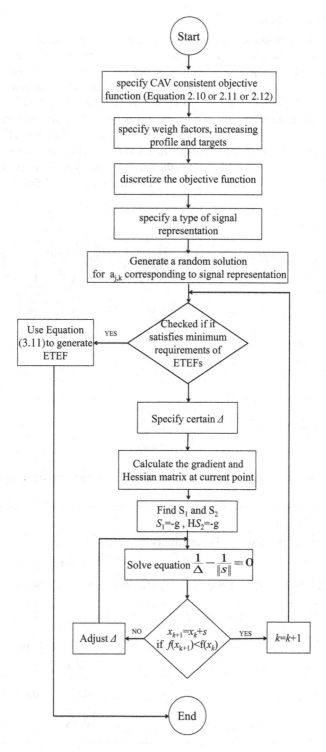

FIGURE 4.6 Algorithm for generating CAV-consistent endurance time excitations.

40-Second Excitations

In this section, long-time CAV-consistent endurance time excitations are generated by application of the proposed method, and the results are presented. As stated before, three types of objective functions are defined. In order to investigate the efficiency of these three objective functions, three series of endurance time excitations are generated. The series "lca", "lcr", and "lc" are generated based on F_{CAV-A}, F_{CAV-R}, and F_{CAV-M} objective functions, respectively. Three excitations are generated for each series. For example, the series "lc" includes {ETA40lc01, ETA40lc02, ETA40lc03} as its members. Now in ETA40lc01, "40" refers to the excitation duration, "lc" refers to the excitation series name, and "01" is the excitation number. Characteristics of the generated ETEFs and corresponding Res_{Total} are summarized in Table 4.2.

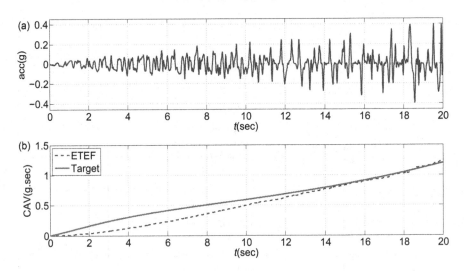

FIGURE 4.7   (a) Generated 20-second ETEF and (b) CAV comparison of ETEF vs. CAV Target.

TABLE 4.2
CAV-Consistent ETEFs Residuals

ETEFs Series	ETEFs Name	Objective Function	Res_{Total} Values	Res_{Total} Average
ETA40lca	01	F_{CAV-A}	0.164	0.169
	02		0.172	
	03		0.172	
ETA40lcr	01	F_{CAV-R}	0.193	0.199
	02		0.201	
	03		0.202	
ETA40lc	01	F_{CAV-M}	0.112	0.114
	02		0.116	
	03		0.115	

The results show that F_{CAV-M} ("lc" series) creates better accuracy. Acceleration time history of ETA40lc01 is illustrated in Figure 4.8. Acceleration spectra of ETA40lc01 at several times are compared with target spectra in Figure 4.9. The investigation regarding displacement spectra and CAV consistency is represented in Figures 4.10 and 4.11. These comparisons show compatibility between endurance time excitations and targets. In Figure 4.11, CAV consistencies of CAV-consistent endurance time excitations and conventional excitation are compared. As asserted by Figure 4.11,

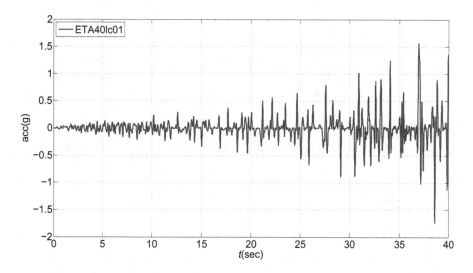

FIGURE 4.8   ETA40lc01 acceleration time history.

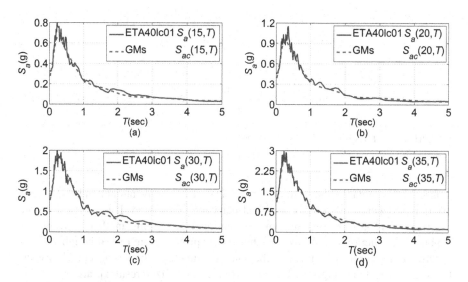

FIGURE 4.9   Comparison between acceleration spectra of ETA40lc01 and target at times: (a) $t = 15$ seconds, (b) $t = 20$ seconds, (c) $t = 30$ seconds, and (d) $t = 35$ seconds.

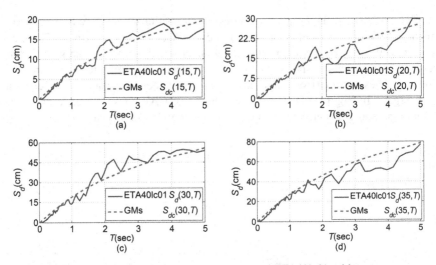

FIGURE 4.10 Comparison of displacement spectra of ETA40lc01 with target spectra at times: (a) $t = 15$ seconds, (b) $t = 20$ seconds, (c) $t = 30$ seconds, and (d) $t = 35$ seconds.

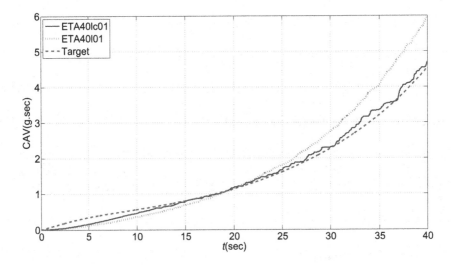

FIGURE 4.11 ETA40lc01 CAV vs. CAV[Target].

at large times, notable discrepancy can be observed in conventional endurance time excitations where CAV consistency had not been considered. It is worth mentioning that ETA40l01 is simulated by conventional generating methods except that exponential function is used as the intensifying profile. Note that although velocity spectra are not directly included in the objective function, there is acceptable consistency, as illustrated in Figure 4.12. Details of the normalized residuals are reported in Table 4.3. The discrepancy in acceleration spectra is about 10%, while displacement spectra and CAV are in turn 16% and 7%.

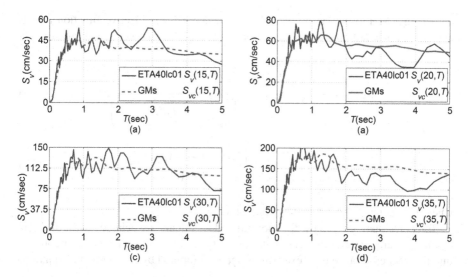

FIGURE 4.12 Comparison of velocity spectra of ETA40lc01 and target at times: (a) $t = 15$ seconds, (b) $t = 20$ seconds, (c) $t = 30$ seconds, and (d) $t = 35$ seconds.

TABLE 4.3

Normalized Residuals of CAV-Consistent ETEFs

ETET Series	ETEF Names	ETEFs Residuals					
		Res_{Sa}		Res_{Su}		Res_{CAV}	
		Values	Average	Values	Average	Values	Average
ETA40lc	01	10.7%	10.5%	16.7%	16.3%	6.3%	7.2%
	02	10.4%		16.1%		7.2%	
	03	10.4%		16.2%		8.1%	

CALCULATION OF RESIDUALS

In order to verify the proposed method and compare the accuracy of generated endurance time excitations therewith, residuals of generated excitations had to be quantified. Residuals are defined as differences between the properties of endurance time excitations and those of the corresponding ground motions. Residuals associated with each dynamic characteristic are generally a function of time and period. Normalized residuals are those averaged over time and period, and thus, no longer a function of time and period. Res_{Sa}, Res_{Su}, and Res_{CAV} are normalized residuals, respectively, associated with acceleration spectra, displacement spectra, and CAV. Res_{Sa} is calculated via Equation (4.6). Substituting $S_a(T,t)$ and $S_{ac}(T,t)$ with $S_u(T,t)$ and $S_{uc}(T,t)$ in Equation (4.6), Res_{Sd} can be calculated. RES_{CAV} is computed using Equation (4.7). These residuals express endurance time excitation accuracy in percentage. Moreover, these residuals do not depend on objective function targets and are applicable for different excitations generated based on different targets.

$$\text{Res}_{S_a} = \frac{1}{t_{max}} \int_0^{t_{max}} \left(\frac{\int_{T_{min}}^{T_{max}} \left| \left(S_a\left(T,t\right) - S_{ac}\left(T,t\right) \right) \right| dT}{\int_{T_{min}}^{T_{max}} S_{ac}\left(T,t\right) dT} \right) dt \tag{4.6}$$

$$\text{Res}_{CAV} = \frac{1}{t_{max}} \int_0^{t_{max}} \left| \frac{CAV\left(t\right) - CAV_C\left(t\right)}{CAV_C\left(t\right)} \right| dt \tag{4.7}$$

where |.| denotes the absolute value operator.

Total residual combines the normalized residuals associated with different dynamic characteristics into one scalar quantity. Total residual is a basis for comparing different endurance time excitations and is computed by the following equation:

$$\text{Res}_{Total} = \frac{[\text{ResRS}_a, \text{ResRS}_d, \text{ResRCAV}].\left[im_{S_a}, im_{S_u}, im_{CAV} \right]}{\sqrt{\left(im_{S_a}\right)^2 + \left(im_{S_u}\right)^2 + \left(im_{CAV}\right)^2}} \tag{4.8}$$

where "." denotes the inner product operator of two vectors. im_{Sa}, im_{Su}, and im_{CAV} are importance factors of acceleration spectra, displacement spectra, and cumulative absolute velocity, respectively. In this chapter, the values of these importance factors are set at one.

REFERENCES

FEMA-p695. (2009). *Quantification of Building Seismic Performance Factors*. Washington, DC: Federal Emergency Management Agency.

Mallat, S. (1989). Multiresolution approximations and wavelet orthonormal bases of $L^2(R)$. *Transaction of American Mathematical Society, 315*, 69–88.

Moré, J. J., & Sorensen, D. C. (1983). Computing a trust region step. *SIAM Journal on Scientific and Statistical Computing, 3*, 553–572. https://doi.org/10.1137/0904038

Newland, D. (1993). *Random Vibrations, Spectral and Wavelet Analysis* (3rd Edition). New York: Longman Scientific & Technical.

5 Nonlinear Analysis-Based Endurance Time Excitation Generation[1]

REVIEW

In this chapter, the simulation procedure of the third generation of endurance time excitations is presented in detail. The objective function of the third generation is discussed in Chapter 2. The procedure of simulating third generation and the resulting excitations are presented. Nonlinear response is directly in producing third generation ETEFs. This hugely increases the computational demand for generation of these series as compared to the second generation of ETEFs. Efficiency of the formulation of the optimization problem is more important in these categories of record generation.

PROCEDURE OF SIMULATING THIRD GENERATION

In this section, different steps in generating endurance time excitations are discussed.

Far-field ground motions which are proposed by FEMA-p695 (FEMAp695, 2009) have been used in this study. These ground motions are recorded from large

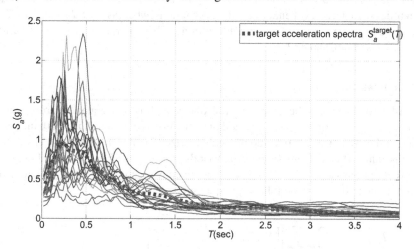

FIGURE 5.1 Target acceleration spectra, $S_a^{target}(T)$.

[1] Chapter source: Mashayekhi, M., Estekanchi, H. E., Vafai, H., & Mirfarhadi, S. A. (2018). Development of hysteretic energy compatible endurance time excitations and its application. *Engineering Structures*, *177*, 753–769.

DOI: 10.1201/9781003216681-5

magnitude events ($M > 6.5$) at sites located greater than or equal to 10 km from fault rupture. This set includes records from soft rock, stiff sites, and shallow crustal sources. Individual records are normalized by their respective peak ground velocities. Their acceleration spectra and median are shown in Figure 5.1. The procedure described in FEMA-p695 (Federal Emergency Management Agency, 2009) is employed in this study. Normalization by peak ground velocity is a simple way to remove unwanted variability between records due to inherent difference in event magnitude, source type, and site condition while maintaining inherent aleatory variability for predicting seismic response assessment.

OBJECTIVE FUNCTION DEFINITION

In the third generation of endurance time excitations, nonlinear displacement residuals are included in the objective function. This objective function is similar to the fifth generation with the difference that hysteretic energy residuals are not taken into account. The objective function of the third generation is given below:

$$
F_{\text{ETEF}}\left(a_g\right) = \int_0^{T_{\max}} \int_0^{t_{\max}} \left\{\left[S_a\left(t,T\right) - S_{aC}\left(t,T\right)\right]^2\right\} dt \, dT
$$

$$
+ \int_1^{\mu_{\max}} \int_0^{T_{\max}} \int_0^{t_{\max}} \left\{\alpha_{u_m}\left[u_m\left(t,T,\mu\right) - u_{mC}\left(t,T,\mu\right)\right]\right\} dt \, dT \, d\mu
$$

(5.1)

where $a_g(t)$ is the acceleration time history of ETEFs, $S_a(t, T)$ and $u_m(t, T, \mu)$ are dynamic characteristics of endurance time excitations, whereas $S_{aC}(t, T)$ and $u_{mC}(t, T, \mu)$ are dynamic characteristics of ground motions. These parameters were explained and discussed in Chapter 2.

DISCRETIZATION

In order to minimize the output of the objective function equation presented in Equation (5.1), nonlinear unconstrained optimization is adopted. It is worth mentioning that the objective function should be discretized in order to be solved. Given that time is sampled at n points (t_j, $j = 1:n$), periods at m points (T_i, $i = 1:m$), and ductility ratios at r points (μ_k, $k = 1:r$), this discretization converts multiple integration to triple summation as shown in the following equation:

$$
F_{\text{ETEF}}\left(a_g\right) = \sum_{i=1}^{m}\sum_{j=1}^{n}\left\{\left[S_a\left(T_i,t_j\right) - S_{aC}\left(T_i,t_j\right)\right]^2\right\} +
$$

$$
\sum_{k=1}^{r}\sum_{i=1}^{m}\sum_{j=1}^{n}\left\{\left[\alpha_{m,\mu_k}\left[u_m\left(t_j,T_i,\mu_k\right) - u_{mc}\left(t_j,T_i,\mu_k\right)\right]+\right]^2\right\}
$$

(5.2)

For simulating third generation, periods (T) are discretized at 120 points between 0.02 and 5 seconds using logarithmic distribution: $T_{min}=0.02$ second and $T_{max}=5$ seconds. The time variable, t, is discretized at 2048 points in intervals of 0.01seconds: $t_{max} = 20.47$ second. The ductility variable is discretized at three ratios (μ): 2, 4, and 8.

SIGNAL REPRESENTATION AND OPTIMIZATION ALGORITHM

Endurance time excitations are represented by their wavelet coefficients. Therefore, in order to generate new excitations, their wavelet coefficients must be specified. In case of a signal consisting of 2^M data points, where M is an integer, discrete wavelet transform requires 2^M wavelet coefficients to fully describe the signal. Discrete wavelet transform decomposes the signal into $M+1$ levels, where the level is denoted as i and the levels are numbered $i=-1, 0, 1, . . , M-1$ (Newland, 1993). For example, a signal with 2048 data points has 2048 wavelet coefficients in ten levels. In order to reduce the optimization variables, the first eight levels are considered and the others are discarded.

Nonlinear unconstrained optimization was used to determine wavelet coefficients. This study used trust-region reflective method as an optimization algorithm, which is a simple yet powerful concept in the field of optimization (Moré & Sorensen, 1983).

RESULTS

One series of endurance time excitations were produced for the third generation. This series is called (kn). The "n" next to the name of series denotes that nonlinear responses have been incorporated in the generating process. In order to form this set, three excitations are generated. {ETA20kn01, ETA20kn02, and ETA20kn03} are members of the (kn) series, where in ETA20kn01, "20" refers to duration, "kn" refers to the type of the series, and "01" is the excitation number in the series. Acceleration time histories of ETA20kn01, ETA20kn02, and ETA20kn03 are shown in Figure 5.2.

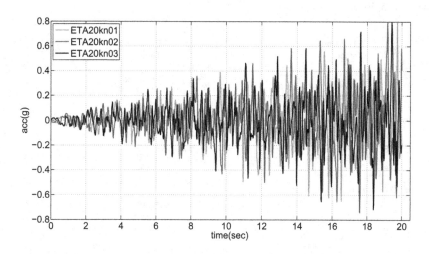

FIGURE 5.2 Acceleration time histories of ETA20kn01, ETA20kn02, and ETA20kn03.

COMPARISON OF DYNAMIC CHARACTERISTICS OF GENERATED ENDURANCE TIME EXCITATIONS AND TARGETS

In this section, dynamic characteristics of ETA20kn01 are presented and compared with targets.

ACCELERATION SPECTRA

Acceptable consistency is demonstrated between ETA20kn01 and targets in Figure 5.3, where the acceleration spectra of ETA20kn01 are compared to the targets at $t=5$, 10, 15, and 20 seconds in turn. Additionally, acceleration spectra of ETA20kn01 are compared with targets for four different periods as shown in Figure 5.4: $T=0.02$, 0.05, 1, and 4 seconds.

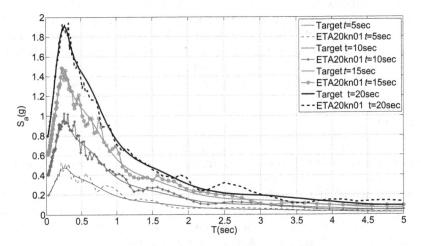

FIGURE 5.3 Comparison between acceleration spectra ETA20kn01 and target acceleration spectra at 5, 10, and 15 seconds.

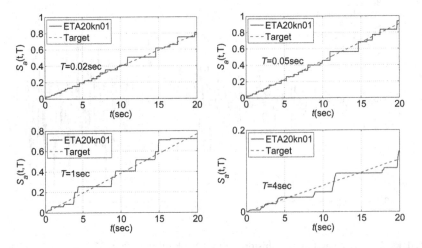

FIGURE 5.4 Comparison of acceleration spectra ETA20kn01 in four different periods: (a) $T=0.02$ second, (b) $T=0.05$ second, (c) $T=1$ second, and (d) $T=4$ seconds.

RESIDUALS OF GENERATED ENDURANCE TIME EXCITATIONS

Residuals of the endurance time excitations are defined as deviations from the targets. These are generally functions of time, ductility ratio, and period. Normalized residual is defined by normalizing and averaging the residuals over periods, ductility ratios, and time. Equation (5.3) calculates normalized residuals for the nonlinear displacement demands. However, the integration over ductility ratio is removed when computing acceleration spectra residuals. The following equations express the accuracy of endurance time excitations in percentage:

$$
\mathrm{NRes}_{u_m} = \frac{1}{(T_{max} - T_{min})(\mu_{max} - 1)} \int_{1}^{\mu_{max}} \int_{T_{min}}^{T_{max}} \frac{\int_{0}^{t_{max}} [u_m(t,T,\mu) - u_{mC}(t,T,\mu)]\,dt}{\int_{0}^{t_{max}} u_{mC}(t,T,\mu)\,dt}\,dT\,d\mu \qquad (5.3)
$$

$$
\mathrm{NRes}_{S_a} = \frac{1}{(T_{max} - T_{min})} \int_{T_{min}}^{T_{max}} \frac{\int_{0}^{t_{max}} [S_a(t,T) - S_{aC}(t,T)]\,dt}{\int_{0}^{t_{max}} S_{aC}(t,T)\,dt}\,dT \qquad (5.4)
$$

Table 5.1 summarizes normalized residuals of (kn) series. Acceleration spectra and nonlinear displacement demonstrate mismatch of 15.6% and 15.1%, respectively.

TABLE 5.1
Residuals of Endurance Time Excitations

		Normalized Residuals			
		NRes_{S_a}		NRes_{u_m}	
ET Series	ETEF Name	Values	Average	Values	Average
kn	ETA20kn01	15.5%	15.6%	14.3%	15.1%
	ETA20kn02	15.8%		15.4%	
	ETA20kn03	15.4%		15.6%	

REFERENCES

FEM Ap695. (2009). *Quantification of Building Seismic Performance Factors, FEMA P-695*. Washington, DC: Federal Emergency Management Agency.

Moré, J. J., & Sorensen, D. C. (1983). Computing a trust region step. *SIAM Journal on Scientific and Statistical Computing*, 3, 553–572. https://doi.org/https://doi.org/10.1137/0904038

Newland, D. (1993). *Random Vibrations, Spectral and Wavelet Analysis* (3rd Edition). New York: Longman Scientific & Technical.

6 Generating ETEFs Considering Spectral Energy Content[1]

REVIEW

In this chapter, a simulation procedure for the fifth generation of endurance time excitations is presented. Typical objective functions of the fifth generation are discussed in Chapter 2. Fifth-generation excitation functions are important from practical viewpoint. They are intended to produce consistent damage levels at matching spectral intensities to ground motions. Generating these ETEFs requires computational effort comparable to third-generation ETEFs. The procedure of simulating fifth generation and the resulting excitations are explained in this chapter.

PROCEDURE OF SIMULATING FIFTH GENERATION

In this section, different steps in the process of generating endurance time excitations are discussed. Far-field ground motions which are proposed by FEMA-p695 (FEMAp695, 2009) have been used in ETEFs explained in this chapter. In order to minimize the output of the objective function Equation (2.21) of Chapter 2 for the fifth generation, nonlinear unconstrained optimization is adopted. It is worth mentioning that the objective function should be discretized in order to be solved. Given that time is sampled at n points ($t_j, j = 1:n$), periods at m points ($T_i, i = 1:m$), and ductility ratios at r points ($\mu_k, k = 1:r$), this discretization converts multiple integration to triple summation as shown in the following equation:

$$F_{\text{ETEF}}(a_g) = \sum_{i=1}^{m} \sum_{j=1}^{n} \left\{ \left[S_a(T_i, t_j) - S_{aC}(T_i, t_j) \right]^2 \right\} \tag{6.1}$$

$$+ \sum_{k=1}^{r} \sum_{i=1}^{m} \sum_{j=1}^{n} \left\{ \begin{bmatrix} \alpha_{m,\mu_k} \left[u_m(t_j, T_i, \mu_k) - u_{mc}(t_j, T_i, \mu_k) \right] + \\ \alpha_{EH,\mu_k} \left[EH(t_j, T_i, \mu_k) - EH_c(t_j, T_i, \mu_k) \right] \end{bmatrix}^2 \right\}$$

For simulating fifth generation, periods (T) are discretized at 120 points between 0.02 and 5 seconds using logarithmic distribution: $T_{\min} = 0.02$ second and $T_{\max} = 5$ seconds.

---

[1] Chapter source: Mashayekhi, M., Estekanchi, H. E., Vafai, H., & Mirfarhadi, S. A. (2018). Development of hysteretic energy compatible endurance time excitations and its application. *Engineering Structures*, *177*, 753–769.

DOI: 10.1201/9781003216681-6

The time variable, t, is discretized at 2048 points in intervals of 0.01 second: $t_{max} = 20.47$ seconds. The ductility variable is discretized at three ratios (μ): 2, 4, and 8.

SIGNAL REPRESENTATION AND OPTIMIZATION ALGORITHM

Endurance time excitations are represented by their wavelet coefficients. Therefore, in order to generate new excitations, their wavelet coefficients must be specified. In case of a signal consisting of 2^M data points, where M is an integer, discrete wavelet transform requires 2^M wavelet coefficients to fully describe the signal. Discrete wavelet transform decomposes the signal into $M+1$ levels, where the level is denoted as i and the levels are numbered $i = -1, 0, 1, \ldots, M-1$ (Newland, 1993). For example, a signal with 2048 data points has 2048 wavelet coefficients in ten levels. In order to reduce the optimization variables, the first eight levels are considered and the others are discarded.

Nonlinear unconstrained optimization was used to determine wavelet coefficients. This study used trust-region reflective method as an optimization algorithm, which is a simple yet powerful concept in the field of optimization (Moré & Sorensen, 1983).

RESULTS

One series of endurance time excitations were produced for the fifth generation. This series is called "kd". The "d" in (kd) denotes that damage compatibility has been considered. Three excitations are generated for kd series. {ETA20kd01, ETA20kd02, and ETA20kd03} are members of the (kd) series, where in ETA20kd01, "20" refers to duration, "kd" refers to the type of the series, and "01" is the excitation number in the series. Optimization convergence histories of ETA20kd01, ETA20kd02, and ETA20kd03 are shown in Figure 6.1. Acceleration time histories of ETA20kd01, ETA20kd02, and ETA20kd03 are shown in Figure 6.2.

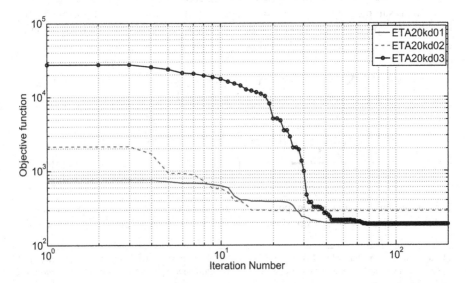

FIGURE 6.1   Optimization convergence histories of simulating of ETA20kd01, ETA20kd02, and ETA20kd03.

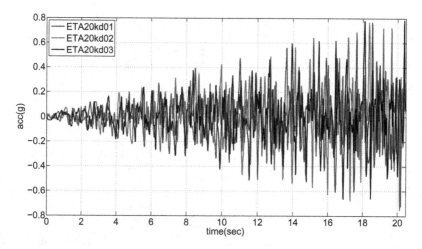

FIGURE 6.2　Acceleration time histories of ETA20kd01, ETA20kd02, and ETA20kd03.

COMPARISON OF DYNAMIC CHARACTERISTICS OF GENERATED ENDURANCE TIME EXCITATIONS AND TARGETS

In this section, dynamic characteristics of ETA20kd01 are presented and compared with targets.

ACCELERATION SPECTRA

Acceptable consistency is demonstrated between ETA20kd01 and targets in Figure 6.3, where the acceleration spectra of ETA20kd01 are compared to the targets at $t = 5$, 10, and 15 seconds in turn. Additionally, acceleration spectra of ETA20kd01 are compared with targets for four different periods as shown in Figure 6.4: $T = 0.02$, 0.05, 1, and 4 seconds.

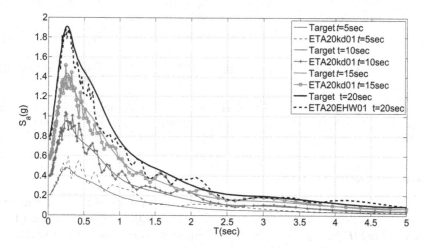

FIGURE 6.3　Comparison between acceleration spectra ETA20kd01 and target acceleration spectra at 5, 10, and 15 seconds.

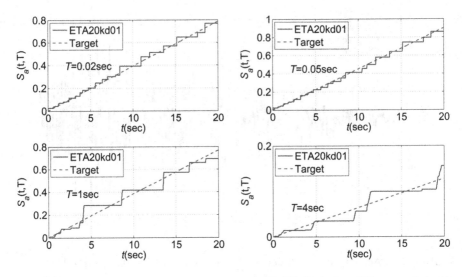

FIGURE 6.4 Comparison of acceleration spectra ETA20kd01 in four different periods: (a) $T=0.02$ second, (b) $T=0.05$ second, (c) $T=1$ second, and (d) $T=4$ seconds.

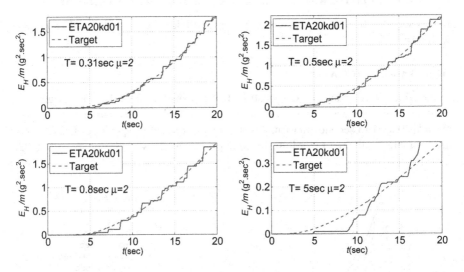

FIGURE 6.5 Comparison between hysteretic energy of ETA20kd01 and targets for $\mu=2$ at four different periods: (a) $T=0.02$ second, (b) $T=0.05$ second, (c) $T=1$ second, and (d) $T=4$ seconds.

HYSTERETIC ENERGY DEMANDS

At different ductility ratios, times, and periods, comparisons were drawn between the hysteretic energy of endurance time excitations ($E_H(t, T, \mu)$) and the target hysteretic energy ($E_H(t, T, \mu)$). Figures 6.5–6.7 report striking consistency between the hysteretic energy of ETA20kd01 and the targets.

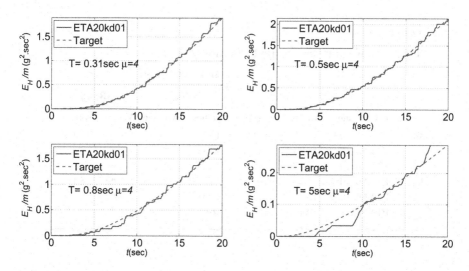

FIGURE 6.6 Comparison between hysteretic energy of ETA20kd01 and targets for $\mu = 4$ at four different periods: (a) $T = 0.02$ second, (b) $T = 0.05$ second, (c) $T = 1$ second, and (d) $T = 4$ seconds.

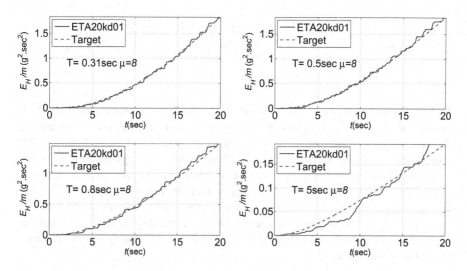

FIGURE 6.7 Comparison between hysteretic energy of ETA20kd01 and targets for ductility ratio of 8 at four different periods: (a) $T = 0.02$ second, (b) $T = 0.05$ second, (c) $T = 1$ second, and (d) $T = 4$ seconds.

RESIDUALS OF GENERATED ENDURANCE TIME EXCITATIONS

As stated in Chapter 5, residuals of the endurance time excitations are defined as deviations from the targets. Normalized residual is defined by normalizing and averaging the residuals over periods, ductility ratios, and time. Equation (6.2) calculates normalized residuals for the hysteretic energy demands. The normalized residuals for

TABLE 6.1

Residuals of Endurance Time Excitations

ET		Normalized Residuals					
		NRes_{s_a}		NRes_{u_m}		NRes_{E_H}	
Series	ETEF Name	Values	Average	Values	Average	Values	Average
kn	ETA20kn01	15.5%	15.6%	14.3%	15.1%	39.2%	41.1%
	ETA20kn02	15.8%		15.4%		41.2%	
	ETA20kn03	15.4%		15.6%		42.9%	
kd	ETA20kd01	9.3%	10.2%	16.4%	16.8%	7.9%	8.1%
	ETA20kd02	10.8%		17.8%		8.9%	
	ETA20kd03	10.4%		16.3%		7.5%	

nonlinear displacement and acceleration spectra are explained in Chapter 5. The following equation expresses the accuracy of endurance time excitations in percentage:

$$\text{NRes}_{E_H} = \frac{1}{(T_{\max} - T_{\min})(\mu_{\max} - 1)} \int_{1}^{\mu_{\max}} \int_{T_{\min}}^{T_{\max}} \frac{\int_{0}^{t_{\max}} [E_H(t,T,\mu) - E_{H,C}(t,T,\mu)]\,dt}{\int_{0}^{t_{\max}} E_{HC}(t,T,\mu)\,dt}\, dT\, d\mu \qquad (6.2)$$

Table 6.1 summarizes normalized residuals of "kd" series. Normalized residuals of "kn" series are also provided for comparison. Acceleration spectra, nonlinear displacement, and hysteretic energy demonstrate mismatch of 10%, 16%, and 8%, respectively. It can be concluded that including hysteretic energy not only increases the compatibility of hysteretic energy but also improves the accuracy of acceleration spectra consistency. Hysteretic energy residual is improved from 41% for "kn" series to about 8% for "kd" series where hysteretic energy consistency is included in the generating process. The 1% increase in nonlinear displacement residuals can be overlooked, considering the improvement in hysteretic energy and acceleration spectra accuracy levels.

DAMAGE SPECTRA OF GENERATED ENDURANCE TIME EXCITATIONS VS. GROUND MOTIONS

In order to evaluate the significance of hysteretic energy compatible endurance time excitations, damage spectra of (kd) series and (kn) series were compared with the counterpart of FEMAp695 records. Damage spectrum is a plot of the damage occurred in a series of single degree-of-freedom (SDOF) systems with different periods which are subjected to the same excitation. Although these SDOF systems have different periods, other dynamic characteristics such as damping ratio and earthquake coefficients are same. Earthquake coefficient is the ratio of the lateral strength capacity of the SDOF system to the SDOF system weight.

Several parameters can be used to represent the damage level of systems. Therefore, several forms of damage spectrum can be defined. In this chapter, a specific variation of the Park–Ang damage index which is suggested by Kunnath, Reinhorn, and Abel (1992) is employed. This damage index which is also known as modified Park–Ang damage index slightly transforms Park–Ang damage index to consider the permanent deformation. Besides that, modified Park–Ang damage index assigns "0" to an undamaged state and "1" to a fully collapsed state. Kunnath's formula is shown in the following equation:

$$D_{PA} = \frac{\theta_m - \theta_r}{\theta_u - \theta_r} + \frac{\beta}{M_y \theta_u} \int dE_h \qquad (6.3)$$

Here, θ represents the member-end rotation. θ_r stands for the recoverable rotation during unloading. M_y is the yield moment capacity, θ_u is the ultimate rotation capacity of the member under monotonic static loading. β is the model parameter (herein considered 0.15), and dE_h is the incremental hysteretic energy demand.

Endurance time excitations and ground motions damage spectra are shown and compared at two hazard levels, return period (RP) equals to 500 and 1000 years in Figures 6.8 and 6.9, respectively. At each level, four earthquake coefficients are considered, namely $C = 0.15$, 0.2, 0.25, and 0.3. Here, ductility capacity (μ_U) which is defined as the ratio of ultimate rotation capacity to recoverable rotation capacity is taken to be 10.

These figures show that the damage spectra of the newly generated endurance time excitations have satisfactory consistency with ground motions. It can be seen

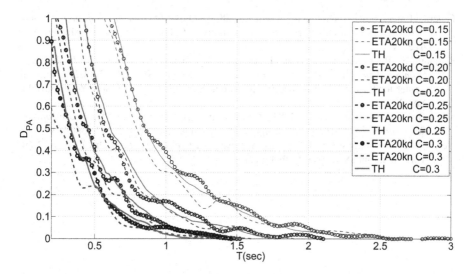

FIGURE 6.8 Comparison between damage spectra of ETA20kd and GM: $\xi = 5\%$, $\mu_u = 10$, and EPP behavior at RP = 500 years.

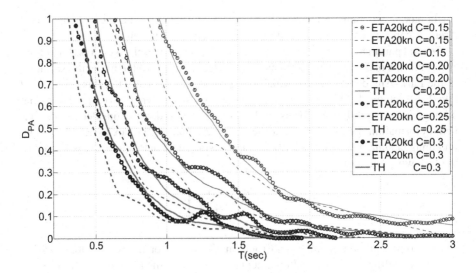

FIGURE 6.9 Comparison between damage spectra of ETA20kd and GM: $\xi = 5\%$, $\mu_u = 10$, and EPP behavior at RP = 1000 years.

that hysteretic compatible excitations (kd series) are considerably more consistent with recorded ground motions as compared to kn series. This is an improvement to the previous endurance time excitations that underestimate damage spectra of ground motions. The underestimation of the previous excitations is more severe at low period region.

REFERENCES

Federal Emergency Management Agency. (2009). *Quantification of Building Seismic Performance Factors, FEMA P-695*. Washington, DC: Federal Emergency Management Agency.

Kunnath, S., Reinhorn, A.., & Abel, J. (1992). *IDARC V3.0: A Program for the Inelastic Damage Analysis of RC Structures*. Technical Report NCEER 92–0022. National Center for Earthquake Engineering Research, State University of New York, Buffalo, New York.

Moré, J. J., & Sorensen, D. C. (1983). Computing a trust region step. *SIAM Journal on Scientific and Statistical Computing, 3*, 553–572. https://doi.org/10.1137/0904038

Newland, D. (1993). *Random Vibrations, Spectral and Wavelet Analysis* (3rd Edition). New York: Longman Scientific & Technical.

7 Application of Meta-Heuristic Optimization Methods in Generating ETEFs[1]

REVIEW

As explained in the previous chapters, numerical optimization is a key tool for producing ETEFs. In this chapter, application of a novel optimization method based on Imperialist Competitive Algorithm (ICA) in simulating Endurance Time (ET) excitations is explained. This method proved to be more effective than standard genetic algorithms for solving the objective function. Unconstrained nonlinear optimization is commonly used to simulate endurance time excitation functions. Simulation of endurance time excitations by using evolutionary algorithms is challenging due to the presence of a large number of decision variables that are highly correlated. The new excitation results were compared with the current practice for simulation of ET excitations. Other optimization methods can also be employed using the framework presented in this chapter.

IMPERIALIST COMPETITIVE ALGORITHM

ICA as an evolutionary algorithm simulates the social–political process of imperialism and imperialist competition. This algorithm contains a population of countries. The pseudo-code of the algorithm is as follows:

Step 1: Generating initial countries
    Similar to other evolutionary algorithms, the ICA starts with an initial population called "country". Each country is identified by a set of decision variables. That is,

$$\text{country}_i = \left[x_i, x_2, ..., x_{N_{var}} \right] \tag{7.1}$$

where N_{var} is the dimension of the optimization problem.

[1] Chapter source: Mashayekhi, M., Estekanchi, H. E., Vafai, H., & Ahmadi, G. (2019). An evolutionary optimization-based approach for simulation of endurance time load functions. *Engineering Optimization*, 51(12), 2069–2088.

DOI: 10.1201/9781003216681-7

The primary locations of countries are determined by assigning a set of values to each decision variable. Because of the dynamic nature of the endurance time excitations simulation problems, the variables are highly correlated. Moreover, the objective function is complicated and requires considerable evaluation time. If the correlation of variables is not considered in the algorithm, the large numbers of variables (in the order of 500) hinder the global convergence. A method is developed to generate initial countries considering the correlation of variables. This method is not limited to this problem and it can be applied to other problems.

This initializing method starts with a set of simulated excitations. It should be mentioned that the endurance time excitations are generated based on different objective functions and the present method only requires the correlation of their variables. This method artificially generates initial countries based on statistics of the existing endurance time excitations. It should be mentioned that none of the simulated excitations are used as initial countries and only their statistics are employed for generating initial countries. Moreover, the existing endurance time excitations are not optimized based on the objective function of this study and their objective functions widely differ from it. Respective variable values of endurance time excitations are summarized in a matrix (Y) as depicted in Table 7.1. The mean vector M_Y and the covariance matrix Σ_{YY} are estimated by using the data in Table 7.1. M_Y is a vector whose element is the mean of each column of Y. Also, the covariance matrix can be computed according to

$$\Sigma_{YY} = D_Y R_{YY} D_Y \qquad (7.2)$$

where D_Y is a diagonal matrix containing standard deviation quantities of each column and R_{YY} is the correlation coefficient matrix. Using the aforementioned statistics, one can infinitely generate artificial samples as initial countries according to the following equation:

$$Z = D_Y L_Y U + M_Y \qquad (7.3)$$

TABLE 7.1

Sample Y Matrix (Data of the Existing Endurance Time Excitations)

Variable Number	x_1	x_2	$\cdots$	$x_{N_{var}}$
1	$x_{1,1}$	$x_{1,2}$	$\cdots$	$x_{1,N_{var}}$
2	$x_{1,2}$	$x_{2,2}$	$\cdots$	$x_{2,N_{var}}$
.	.	.	.	.
.	.	.	.	.
.	.	.	.	.
M	$x_{M,1}$	$x_{M,,2}$	$\cdots$	$x_{M,N_{var}}$

where L_Y is the lower triangular decomposition of the correlation matrix R_{YY} so that $R_{YY} = L_Y L_Y^T$. Because the correlation matrix is positive definite, L_Y can be computed by the Cholesky decomposition of R_{YY}. U is the standard normal random vector with zero mean and unit covariance matrix.

Step 2: Generating initial empires

The best countries in the initial population are selected as imperialists, while others are considered as their relevant colonies. The cost of each country identifies its power. The total number of initial countries is set to $N_{country}$ and the number of the most powerful countries, which act as imperialists for empires, is equal to N_{imp}. The remaining N_{col} ($N_{col} = N_{country} - N_{imp}$) of the initial countries is the number of colonies of these empires. All colonies of initial countries are divided among imperialists based on their power. Thus, the number of colonies of an empire must be inversely proportional to its cost value. In order to proportionally divide colonies among imperialists, a normalized cost for an imperialist is defined as follows:

$$C_j = e^{\frac{-\alpha f_{cost}^{(imp,j)}}{\max\left(f_{cost}^{(imp,j)}\right)}} \tag{7.4}$$

where $f_{cost}^{(imp,j)}$ is the cost of the j-th imperialist, α is a weight constant, and C_j is the normalized cost. Colonies are divided among empires based on their power and for the j-th empire will be as follows:

$$NC_j = \text{round}\left(\frac{C_j}{\sum\limits_{j=1}^{N_{imp}} C_j} \times N_{col}\right) \tag{7.5}$$

where NC_j is the number of colonies associated with the j-th empire, which are selected randomly among empires. These colonies together with the j-th imperialist form the j-th empire.

Step 3: Moving colonies toward the relevant imperialist

In the ICA, all colonies are moved toward imperialists inspired by the assimilation policy pursued by some of the former imperialist states. This movement is shown in Figure 7.1. In this movement, the colony moves toward its imperialist by a random vector that is uniformly distributed between 0 and 1. This movement is given as follows:

$$\{x\}_{new} = \{x\}_{old} + \beta \times d \times \{\text{rand}\} \otimes \{V_1\} \tag{7.6}$$

where β is a parameter that controls the amplitude of movement and d is the distance between colony and imperialist. $\beta > 1$ makes the colonies move closer to the imperialist state from both sides. A value of β very close to one reduces the search power of the algorithm. In Equation (7.12), $\{V_I\}$ is a unit vector with its start point being the previous location of the colony and with its direction being toward the imperialist location. {rand} is a random vector and the sign $\otimes$ denotes element-by-element multiplication. Since these random numbers are not necessarily equal, the colony deviates from the direction of $\{V_I\}$.

Step 4: Revolution

In the ICA, an operation, which is called revolution, brings sudden random changes in the position of some colonies in the search space. Revolution in the ICA is the counterpart of mutation in GA, which increases exploration and prevents the early convergence of countries to the local optimum. Figure 7.2 illustrates the role of the revolution in the ICA. The colony located at the point A can be transferred to some point between A and B by assimilation. In this case, the colony is trapped at the local optimum. On the other hand, through a revolution, the colony suddenly jumps from point A to point C, resulting in an escape from local optimum.

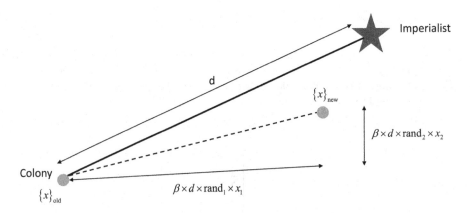

FIGURE 7.1   Movement of colony toward the imperialist.

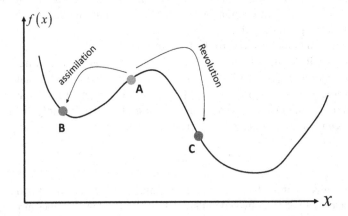

FIGURE 7.2   Illustration of revolution in the ICA.

In the revolution operation, μ_{rev} denotes the percentage of colonies which are chosen for this operation. In fact, the number of rounds ($\mu_{rev}*N_{col}$) is randomly chosen from colonies of an empire for suddenly changing their positions.

In the revolution operation, p_{rev} denotes the percentage of the countries' variables which are changed during the revolution. Consequently, at a problem with N_{var} variables, $N_{rev}=$round ($p_{rev}*N_{var}$) variables are changed. A number of N_{rev} variables are selected randomly from the total number of N_{var}. Indices of these selected variables are referred to as $\{i_1, i_2,..., i_{N_{rev}}\}$. The revolution operation is performed according to the following equation:

$$\{x_i\}_{new} = \begin{cases} \{x_i\}_{old} + \sigma_{rev} \times \text{rand} & i \in \{i_1,i_2,...,i_{N_{rev}}\} \\ \{x\}_{old} & i \notin \{i_1,i_2,...,i_{N_{rev}}\} \end{cases} \tag{7.7}$$

where σ_{rev} is the standard deviation of revolution and rand is a random number, which is normally distributed between 0 and 1. The above equation for revolution is designed for continuous problems and differs from those used in discrete optimization problems (Rabiee et al., 2014; Jolai et al., 2012; and Wang et al., 2014).The value of σ is derived according to the following equation:

$$\sigma_{rev} = 0.1\left(x_{i,max} - x_{i,min}\right) \tag{7.8}$$

where $x_{i,\ max}$ and $x_{i,\ min}$ are upper bound and lower bound of optimization variables, respectively.

Step 5: Revolution of imperialists

The revolution operation is also performed on imperialists. All formulations and assumptions for imperialist evolutions are in total correspondence to those for the colonies, with the difference that with the imperialists, a revolution is not accepted unless the cost function of the new point is improved.

Step 6: Imperialist updating

With assimilation and revolution operations, the cost of new countries is changed. The cost function of new colonies is compared with the cost function of the updated imperialist in each empire. If the new position of a colony is better than that of its relevant imperialist, the imperialist and the colony exchange places, and the country with the lower cost becomes the imperialist.

Step 7: Total power of an empire

The total power of an empire is computed based on the power of its imperialist and a fraction of the power of its colonies. That is,

$$TC_j = f_{cost}^{(imp,j)} + \xi \frac{\sum_{i=1}^{NC_j} f_{cost}^{(col,i)}}{NC_j} \tag{7.9}$$

where TC_j is the total cost of the j-th empire and ξ is a positive number, which is considered to be less than 1. The value of 0.1 for ξ is found to be suitable in most of the implementations. A small value of ξ emphasizes a greater influence on the imperialist power in determining the total power of an empire, while a large value of ξ indicates larger influence of the mean power of colonies in determining the total power of the empire.

Step 8: Imperialist competition

Imperialist competition is another strategy utilized in the ICA. During the competition among imperialists, weaker empires will gradually collapse. Weaker countries lose their colonies, while stronger empires take control of the colonies of the weaker empires. The imperialist competition is modeled by picking some of the weakest colonies of the weakest empire and imposing a competition on all other empires to take control of that colony. In fact, the weakest colony from the weakest empire is the subject of competition among $(N_{\text{imp}}-1)$ empires.

Empires are ranked based on their total power. The first empire, i.e., the weakest one, loses the weakest colony and the empires of ranks 2 to N_{imp} compete to take possession of that colony. The normalized total power is defined by

$$NTC_j = e^{\dfrac{-\alpha C_j}{\max\limits_{i=2:N_{\text{imp}}} (C_i)}} \tag{7.10}$$

where NTC_j is the normalized total power of the j-th empire. The possession probability of each empire is computed based on its normalized total power according to

$$P_j = \left| \dfrac{NTC_j}{\sum\limits_{i=2}^{N_{\text{imp}}} NTC_i} \right| \tag{7.11}$$

A mechanism similar to the Roulette Wheel in GA is utilized to allot the weakest colony among empires based on their possession probabilities.

Step 9: Collapsing the weakest empire (if necessary)

When an empire loses all of its colonies, it is assumed to collapse. In the current implementation, the imperialist relevant to the collapsed empire is allotted to other empires based on the imperialist competition approach discussed in the previous step.

Step 10: Terminating criteria control

There are several terminating criteria, which can be adopted to control termination of the algorithm. For example, when the amount of improvement in the best result is reached, the algorithm is stopped. In the current implementation, the number of iterations is checked, and if it reaches a pre-specified value, the searching process is stopped. The movement of colonies toward their relevant imperialists along the imperialist competition makes all countries to converge on a state in which there is one empire in the world and other countries are colonies of the imperialist of that empire.

A summary of the ICA which is implemented in this study is depicted in Figure 7.3.

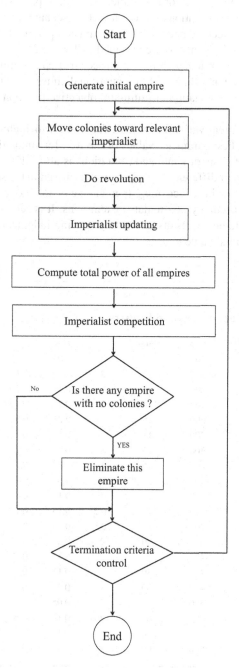

FIGURE 7.3   The Imperialist Competitive Algorithm (ICA).

PARAMETER TUNING OF ICA

In this section, the proposed method is applied to simulate new endurance time excitations. First, it is required to find optimal parameter values of the ICA, i.e., n_{pop}, n_{emp}, β, μ_{rev}, ζ, and p_{rev}. In order to find the optimal parameter values for simulating endurance time excitations, 20 optimization scenarios are defined. For each scenario, the ICA is executed and the best cost and computational time are evaluated. Characteristics of these scenarios are provided in Table 7.2.

The results of the mentioned optimization scenarios are summarized in Table 7.3, where the cost function values and computational times are provided. The mean value and the standard deviation of minimum objective function values are 894 and 105, respectively.

Table 7.3 shows great variability in cost functions which emphasizes the necessity of determining best parameter values of the ICA for simulating endurance time excitations. COV of costs of simulated excitations is about 12%, which is relatively high. The influences of different parameters are discussed in the subsequent sections.

Figure 7.4 shows the influence of μ_{rev} (percentage of colonies chosen for revolution in the ICA) on the accuracy of simulated excitations. It is seen that μ_{rev}s of 0.1 and 0.15 generally create better results. As an engineering judgment, the value of 0.1 is assigned as the optimal value to μ_{rev}.

TABLE 7.2
Parameters of Defined Scenarios

Scenario ID	n_{pop}	n_{emp}	β	μ_{rev}	ζ	p_{rev}
SC1	200	15	2	0.05	0.1	0.1
SC2	200	15	2	0.1	0.1	0.1
SC3	200	15	2	0.15	0.1	0.1
SC4	200	15	2	0.2	0.1	0.1
SC5	200	15	1	0.05	0.1	0.1
SC6	200	15	1	0.1	0.1	0.1
SC7	200	15	1	0.15	0.1	0.1
SC8	200	15	1	0.2	0.1	0.1
SC9	200	15	0.5	0.05	0.1	0.1
SC10	200	15	0.5	0.1	0.1	0.1
SC11	200	15	0.5	0.15	0.1	0.1
SC12	200	15	0.5	0.2	0.1	0.1
SC13	400	30	2	0.05	0.1	0.1
SC14	400	30	2	0.1	0.1	0.1
SC15	400	30	2	0.15	0.1	0.1
SC16	400	30	2	0.2	0.1	0.1
SC17	800	60	2	0.05	0.1	0.1
SC18	800	60	2	0.1	0.1	0.1
SC19	800	60	2	0.15	0.1	0.1
SC20	800	60	2	0.2	0.1	0.1

TABLE 7.3

The Minimum Objective Functions of Defined Scenarios

Scenario ID	Minimum Objective Function Values	Computational Time (seconds)
SC1	908.8	342,219
SC2	839.7	343,619
SC3	836.3	365,175
SC4	842.6	333,612
SC5	965.5	345,001
SC6	875.6	341,048
SC7	907.7	329,589
SC8	974.1	318,502
SC9	1076.4	337,265
SC10	1094.6	346,044
SC11	1029	329,251
SC12	1065	320,830
SC13	870.1	679,075
SC14	816.7	669,614
SC15	818.1	671,705
SC16	739.0	723,261
SC17	810.1	885,964
SC18	789.0	984,483
SC19	800.7	997,081
SC20	815.4	988,556

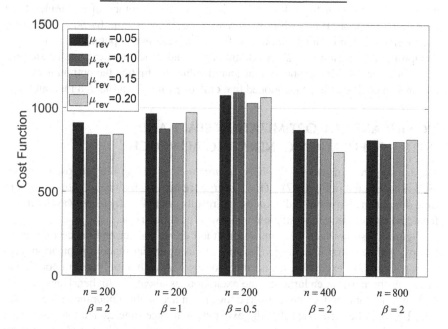

FIGURE 7.4 Variations of the best costs of simulated endurance time excitations for different μ_{rev}.

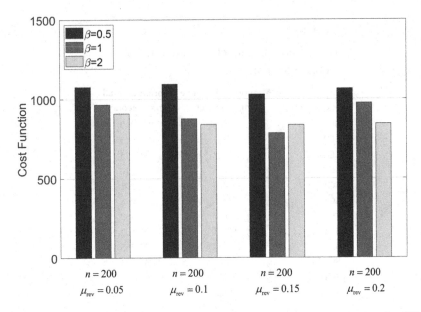

FIGURE 7.5 Variations of the best costs of simulated endurance time excitations for different values of β.

Figure 7.5 shows the influence of β (the parameter that controls the amplitude of movement) on the accuracy of simulated endurance time excitations. It can be seen that $\beta = 2$ leads to more accurate endurance time excitations in all cases.

Considering the trade-off between accuracy and computational time, the population size of 400 is used as the optimal number. It is observed that an increase in the number of population from 400 to 800 improves the accuracy of endurance time excitations by about 0.7%, while it increases the analysis time by 50%. The 0.7% improvement is too infinitesimal to justify the increase in computational time.

Optimal ICA parameter values for simulating endurance time excitations are provided in Table 7.4. The number of considered values for finding the optimum value is also presented. It should be mentioned that only one value is considered for ζ and p_{rev}.

SIGNIFICANCE OF OPTIMIZATION SPACE AND INITIAL POPULATION GENERATING APPROACH

The convergence history of the ICA optimization by using the proposed method (SC14) is depicted in Figure 7.6. The convergence history associated with the ICA in time domain is also presented. In addition, results of the ICA by using wavelet transform and random initial population are shown for comparison.

Figure 7.6 shows that the proposed method creates better endurance time excitations. Although the ICA in wavelet domain, using initial random population, creates an acceptable excitation, the accuracy of the ICA with initial random population based on the existing endurance time excitations is about 5% higher. However, the ICA in time domain does not converge toward an acceptable endurance time excitation. It should be mentioned that the existing endurance time excitations have been simulated in time domain using gradient-based optimization methods and not with the

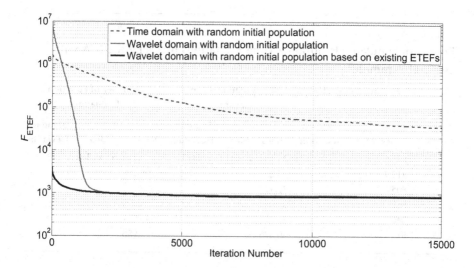

FIGURE 7.6 Convergence history of simulating linear endurance time excitations.

ICA method. Therefore, the fact that the algorithm does not converge in time-domain space using the ICA method is not related to the existing methods of simulating endurance time excitations. It can be seen that the best cost of the time-domain optimization is worse than that in the proposed method, even at initial population.

SIGNIFICANCE OF OPTIMAL PARAMETER IN SIMULATING ENDURANCE TIME EXCITATIONS

In this section, the results of ICA with optimal parameters are compared with results presented in the previous section. Six endurance time excitations are simulated by the ICA using the optimal parameter values listed in Table 7.4. The corresponding results are provided in Table 7.5.

TABLE 7.4

Optimal Parameters of ICA for Simulating Endurance Time Excitations

Parameter	Optimum Value	Number of Considered Values
n_{pop}	400	3
n_{emp}	30	3
β	2	3
μ_{rev}	0.1	4
ζ	0.1	1
p_{rev}	0.1	1

TABLE 7.5

Results of Simulated Endurance Time Excitations by Using the ICA with Optimal Parameters

Run Number	Minimum Objective Function			Computational Time (seconds)			Number of Function Evaluations		
	Values	Mean	Standard Deviation	Values	Mean	Standard Deviation	Values	Mean	Standard Deviation
1	839.6	832	24	723,107	728,519	4433	6,598,242	6,596,363	1283
2	812.7			732,251			6,595,368		
3	805.4			722,575			6,595,989		
4	869.4			731,030			6,594,670		
5	846.9			731,478			6,596,921		
6	816.7			730,676			6,596,986		

From Table 7.5, it can be seen that the simulation of endurance time excitations by the ICA using optimal parameter values leads to 10% improvement and 76% decrease in standard deviation of cost function values. Using optimal parameter values of ICA decreases the mean value of minimum objective function from 894 to 832. In addition, using optimal parameter values decreases the standard deviation from 105 to 24. It is obvious that a higher standard deviation of cost function values requires a larger number of ICA runs to achieve an acceptable endurance time excitation. Both these mentioned facts prove the necessity of using optimal parameter values in simulating endurance time excitations.

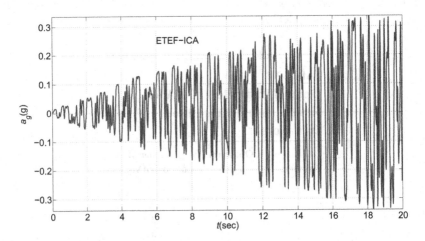

FIGURE 7.7   Acceleration time history of ETEF-ICA.

COMPARISON OF SIMULATED ENDURANCE TIME EXCITATIONS WITH TARGETS

In this section, comparison is drawn between acceleration spectra of simulated endurance time excitation by using ICA with targets. The acceleration time history of an endurance time excitation simulated by the method is shown in Figure 7.7. This excitation is hereafter denoted by ETEF-ICA.

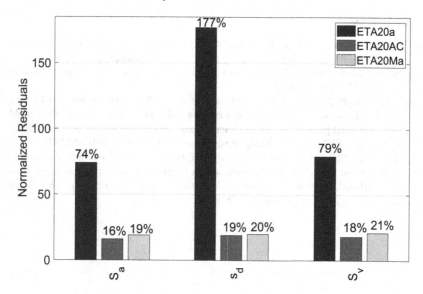

FIGURE 7.8 Acceleration response spectra of simulated linear endurance time excitation vs. targets at (a) $t=5$ seconds, (b) $t=10$ seconds, (c) $t=15$ seconds, and (d) $t=20$ seconds.

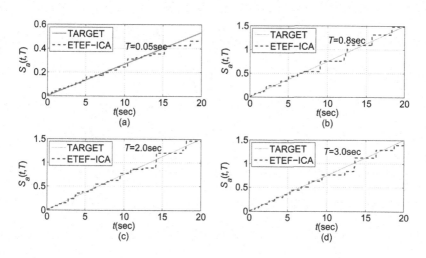

FIGURE 7.9 Time variation of acceleration response spectra of simulated linear endurance time excitation vs. targets for different natural periods of (a) $T=0.05$ second, (b) $T=0.8$ second, (c) $T=2.0$ seconds, and (d) $T=3.0$ seconds.

Acceleration spectra of ETEF-ICA are compared with targets at $t = 5$, 10, 15, and 20 seconds in Figure 7.8. The time variation of the acceleration response spectra is also compared with targets for different periods of $T = 0.05$, 0.8, 2.0, and 3.0 seconds in Figure 7.9. These figures show outstanding correspondence between the simulated endurance time excitation and targets, a correspondence that highlights the efficiency of the proposed method in simulating endurance time excitations.

REFERENCES

Jolai, F., Rabiee, M., & Asefi, H. (2012). A novel hybrid meta-heuristic algorithm for a no-wait flexible flow shop scheduling problem with sequence dependent setup times. *International Journal of Production Research*, *50*(24), 7447–7466. https://doi.org/10.1 080/00207543.2011.653012

Rabiee, M., Sadeghi Rad, R., Mazinani, M., & Shafaei, R. (2014). An intelligent hybrid meta-heuristic for solving a case of no-wait two-stage flexible flow shop scheduling problem with unrelated parallel machines. *International Journal of Advanced Manufacturing Technology*, *71*(5–8), 1229–1245. https://doi.org/10.1007/s00170-013-5375-1

Wang, B., Guan, Z., Li, D., Zhang, C., & Chen, L. (2014). Two-sided assembly line balancing with operator number and task constraints: A hybrid imperialist competitive algorithm. *International Journal of Advanced Manufacturing Technology*, *74*(5–8), 791–805. https://doi.org/10.1007/s00170-014-5816-5

8 Modifying ETEFs as an Alternative to Their Generation[1]

REVIEW

Two methods for simulation of ETEFs) were mentioned in the previous chapters. The first one was based on random vibration theory. This method cannot be employed in practical use in its simple form. The second method utilizes optimization process to simulate excitations. Although the latter method results are satisfactory, they require high computational demand. In this chapter, a new method for producing new endurance time excitations based on the previous ones is explained. The idea is to modify the existing excitations to be compatible with a desired acceleration spectrum. This process, called spectral matching, obviates the need for using the complicated optimization procedures in simulating new excitations. A Fourier-based method for endurance time excitation spectral matching is proposed. This algorithm is then applied in a case study. Alternative methods can be developed based on the idea of spectral matching as explained in this chapter.

ENDURANCE TIME EXCITATIONS SPECTRAL MATCHING

Endurance time excitations spectral matching is a method to modify the existing excitations to produce a desired target spectrum. This method is based on changing frequency amplitudes of current excitations which are denoted as $a_0(t)$ and are represented in frequency domain using Fourier transform as stated in the following equation:

$$A_0(\omega) = \int_{-\infty}^{\infty} a_0(t)e^{i\omega t}\, dt \tag{8.1}$$

---

[1] Chapter source: Mashayekhi, M., Estekanchi, H. E., & Vafai, H. (2020). A method for matching response spectra of endurance time excitations via the Fourier transform. *Earthquake Engineering and Engineering Vibration*, 19(3), 637–648.

DOI: 10.1201/9781003216681-8

Having represented the excitations by the Fourier transform, new excitation time histories $a_1(t)$ are generated by adjusting the Fourier transform coefficients as in the following equation:

$$A_1(\omega) = R(\omega) A_0(\omega) + \Delta\omega \tag{8.2}$$

where $A_1(\omega)$ is the modified endurance time excitation Fourier transform coefficient at frequency ω. These coefficients are obtained by using the multiplicative function $R(\omega)$ and an additive function $\Delta\omega$.

Endurance time excitations spectral matching considers consistency at all intensity levels which correspond to different time windows. It differs from ground motions spectral matching which considers only the entire duration of motions.

In order to adjust the Fourier coefficients of an endurance time excitation, spectral discrepancies are first quantified. Two error functions are defined, both of which express discrepancies as functions of time and frequency. The first error function is defined in the following equation:

$$E(t,\omega) = S_a(t,\omega) - S_{aC}(t,\omega) \tag{8.3}$$

where $S_a(\omega, t)$ is the endurance time excitation acceleration spectra at time t and frequency ω, $S_{aC}(\omega, t)$ is the endurance time excitation target acceleration spectra at time t and frequency ω, and $E(\omega, t)$ is the first error function value quantifying differences between endurance time excitation acceleration spectra and targets.

The second error function $R(\omega, t)$ is defined by the following equation:

$$R(t,\omega) = S_a(t,\omega) / S_{aC}(t,\omega) \tag{8.4}$$

As seen in Equation (8.4), the second error function is a function of frequency and time, and hence, cannot be used in Equation (8.2) directly. This is because the Fourier transform decomposes signals into several harmonic functions with constant amplitudes in signal lengths. In fact, the Fourier amplitudes are constant along time and cannot be a function of time. The denominator of the second error function is zero at time zero, and therefore, the function is indeterminate at time zero. The second error function is only calculated at a time greater than zero. In fact, calculating the error at the initial of endurance time excitations is meaningless.

The multiplicative function which is applicable in Equation (8.2) must be only a function of ω. Two modified multiplicative functions are introduced. The first modified multiplicative function is presented as in Equation (8.5). The area under acceleration spectra along the signal duration is utilized to remove the time dependence. It is worth mentioning that the denominator of this multiplicative function is the area under the target acceleration spectrum; hence, division by zero does not happen in this type of multiplicative function:

$$R_1(\omega) = \frac{\int_0^{t_{max}} S_a(t,\omega)dt}{\int_0^{t_{max}} S_{aC}(t,\omega)dt} = \frac{2t_{target} * \int_0^{t_{max}} S_a(\omega,t)dt}{t_{max}^2 * S_a^T(\omega)} \tag{8.5}$$

The second modified multiplicative function can be determined via Equation (8.6). In the second approach, $S_a(\omega, t)$ is represented as a linear function of time at different frequencies. This representation is presented in Equation (8.7). The denominator of $R_2(\omega)$ is the derivative of $S_{aC}(t, T)$ with respect to t which is $S_a^{target}(\omega)/t_{target}$. The latter value is never zero and this fact prevents the numerical problem in the simulation process.

$$R_2(\omega) = \frac{a_1(\omega)}{S_a^{target}(\omega)/t_{target}} = \frac{t_{target} \times a_1(\omega)}{S_a^{target}(\omega)} \tag{8.6}$$

$$S_a(t,\omega) = a_1(\omega)t + a_0(\omega) \tag{8.7}$$

where $a_1(\omega)$ and $a_0(\omega)$ are linear regression coefficients.

Total error at each frequency can be derived from the first error function $E(\omega,t)$:

$$E(\omega) = \int_0^{t_{max}} E(\omega,t)dt = \int_0^{t_{max}} (S_a(t,\omega) - S_{aC}(t,\omega))dt \tag{8.8}$$

Here, the additive function $\Delta\omega$ is not considered in adjusting the Fourier transform coefficients and only the modified multiplicative function $R(\omega)$ is considered. Not considering the additive function implies that there are no phase modifications in the simulation process.

When $|E(\omega)| \geq E_L$, ETEF time history is filtered by scaling both the real and imaginary parts of the Fourier transform according to Equation (8.9). E_L is a threshold where if the total error defined in Equation (8.8) is greater than E_L, the modification of transformation coefficients is performed. Otherwise, there is no need for any modification. In fact, this logical condition can be considered as a convergence criterion for the simulation process. The value of E_L is computed in accordance with Equation (8.10). To avoid numerical problems, multiplicative coefficients are averaged in specified frequency bands. The same value is considered for all frequencies of a frequency band. It should be mentioned that the denominator which is the

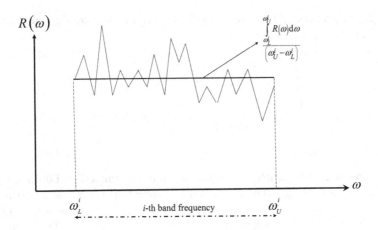

FIGURE 8.1   Band frequencies for computing multiplicative function.

integral of positive numbers over a frequency band can never be zero. This procedure is shown in Figure 8.1.

$$A_1\left(\omega_k\right) = \frac{\left(\omega_U^i - \omega_L^i\right)}{\displaystyle\int_{\omega_L^i}^{\omega_U^i} R(\omega)\,d\omega}\, A_0\left(\omega_k\right) \quad \omega_L^i \leq \omega_k \leq \omega_U^i \tag{8.9}$$

$$E_L = 0.1 \times S_a^{\text{Target}}\left(\omega\right) \times t_{\max} \tag{8.10}$$

where ω_L^i and ω_U^i are the lower and the upper frequency limits of the i-th frequency band.

Utilizing the new Fourier transform coefficients $A_1(\omega)$, a new endurance time excitation time history can be simulated by employing the inverse Fourier transform. It should be considered that the proposed method only changes the frequency amplitude and the phase remains unchanged.

$$a_1\left(t\right) = \frac{1}{2\pi}\int_{-\infty}^{+\infty} A_1\left(\omega\right)e^{-i\omega t}\,d\omega \tag{8.11}$$

Since changing the Fourier transform amplitude at a frequency band may influence other frequency band responses, simulating new endurance time excitations by this method must be iterated several times until convergence is reached. This iterative procedure algorithm is depicted in Figure 8.2.

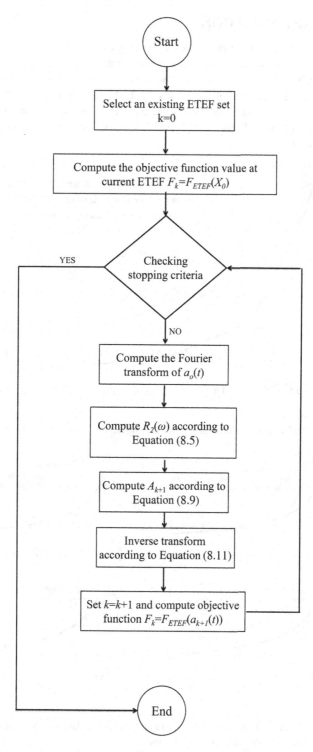

FIGURE 8.2 Algorithm for spectra matching of ETEFs.

STEP-BY-STEP APPLICATION

The proposed method in the previous section is applied in a case study. In this case study, the "a" series is matched with an average spectrum of FEMAp695 records. The "a" series has been generated based on the Iranian National Building Code (BHRC, 2005). The target spectrum of "a" series vs. FEMAp695 is shown in Figure 8.3. ETA20a01 acceleration spectra are compared with target spectra at different time windows as depicted in Figure 8.4. From this figure, discrepancies between these spectra are obvious.

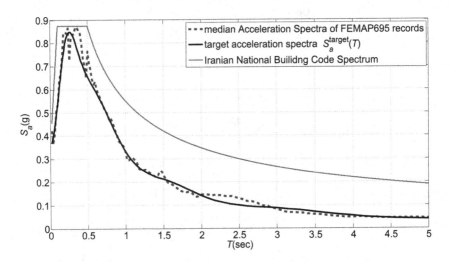

FIGURE 8.3   "a" series acceleration spectra vs. target.

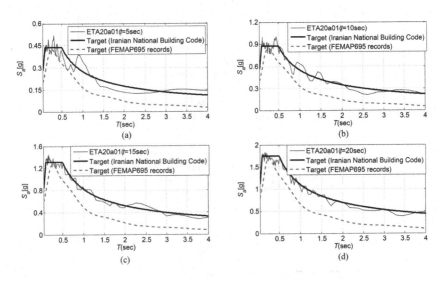

FIGURE 8.4   ETA20a01 acceleration spectra vs. targets at times: (a) $t = 5$ seconds, (b) $t = 10$ seconds, (c) $t = 15$ seconds, and (d) $t = 20$ seconds.

The proposed algorithm shown in Figure 8.2 is applied for ETA20a01. The frequency bands considered are mentioned in Table 8.1. The convergence history is depicted in Figure 8.5. In this figure, two multiplicative functions $R_1(\omega)$ and $R_2(\omega)$ are compared. Whether multiplicative functions are calculated over frequency bands or not is considered as well.

Figure 8.5 shows that the algorithm diverges, unless multiplicative functions are computed over various frequency bands. Result comparison shows that $R_2(\omega)$ brings about more accurate ETEFs. The simulated time history is depicted in Figure 8.6.

TABLE 8.1

Frequency Bands Used in the Spectral Matching

Row Number	Frequency Band (Hz)	Row Number	Frequency Band (Hz)
1	[0.00,0.34]	10	[2.25,2.50]
2	[0.34,0.54]	11	[2.50,2.83]
3	[0.54,0.68]	12	[2.83,3.27]
4	[0.68,0.83]	13	[3.27,4.98]
5	[0.83,1.02]	14	[4.98,19.50]
6	[1.02,1.71]	15	[19.50,24.36]
7	[1.71,1.90]	16	[24.36,29.25]
8	[1.90,2.00]	17	[29.25,39.01]
9	[2.00,2.25]	18	[39.01,50.00]

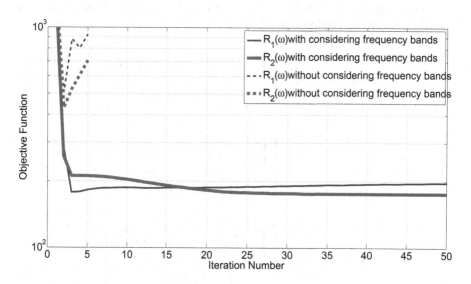

FIGURE 8.5 Convergence history of simulating of ETA20Ma01.

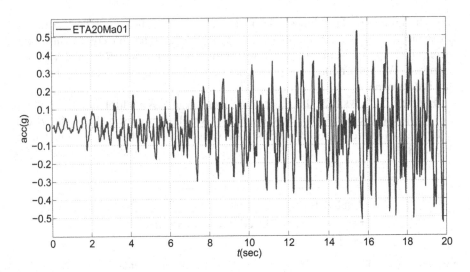

FIGURE 8.6   ETA20Ma01 acceleration time history.

ACCURACY OF SPECTRAL MATCHED ENDURANCE TIME EXCITATIONS

In order to demonstrate the applicability of the proposed algorithm, the matched endurance time excitations accuracy is investigated. In this regard, the Normalized Relative Residual (NRR) is introduced in Equation (8.12). This parameter expresses the endurance time excitations accuracy in percentage, which makes comparison much easier. NRR is computed for acceleration, displacement, and velocity spectra as in Equations (8.12)–(8.14). It should be mentioned that the denominators of these equations are always positive and non-zero.

$$\text{NRR}_{S_a} = \frac{1}{t_{max}} \int_0^{t_{max}} \left(\frac{\int_{T_{min}}^{T_{max}} \left|\left(S_a\left(t,T\right) - S_{ac}\left(t,T\right)\right)\right| dT}{\int_{T_{min}}^{T_{max}} S_{ac}\left(t,T\right) dT} \right) dt \qquad (8.12)$$

$$\text{NRR}_{S_d} = \frac{1}{t_{max}} \int_0^{t_{max}} \left(\frac{\int_{T_{min}}^{T_{max}} \left|\left(S_d\left(t,T\right) - S_{dC}\left(t,T\right)\right)\right| dT}{\int_{T_{min}}^{T_{max}} S_{dC}\left(t,T\right) dT} \right) dt \qquad (8.13)$$

$$\text{NRR}_{S_v} = \frac{1}{t_{max}} \int_0^{t_{max}} \left(\frac{\int_{T_{min}}^{T_{max}} \left|\left(S_v\left(t,T\right) - S_{vC}\left(t,T\right)\right)\right| dT}{\int_{T_{min}}^{T_{max}} S_{vC}\left(t,T\right) dT} \right) dt \qquad (8.14)$$

where $S_{dC}(t, T)$ and $S_{vC}(t, T)$ denote, respectively, ETEFs target displacement and velocity spectra at time t and period T.

The "a" series includes three ETEFs, namely ETA20a01, ETA20a02, and ETA20a03. These ETEFs are given as input motions of the proposed algorithm. Simulated excitations are called ETA20Ma01, ETA20Ma02, and ETA20Ma03. The letter "M" denotes modified endurance time excitations. In order to put the proposed method's efficiency to comparison, three excitations are simulated by conventional simulating approaches. These excitations are called ETA20AC01, ETA20AC02, and ETA20AC03. Table 8.2 summarizes the NRR of these ETEFs.

According to Table 8.2, the proposed method decreases the NRR from 73% for ETA20a01 acceleration spectra to 17% for the spectral matched ETEFs. Figure 8.7 illustrates the average NRR of ETA20a, ETA20Ma, and ETA20AC series.

TABLE 8.2

Comparison of Accuracy of Matched ETA20a with Accuracy of ETA20AC

		Normalized Residuals		
ETEFs Series	ETEFs Number	NRR_{S_a}	NRR_{S_d}	NRR_{S_v}
ETA20a	01	73%	171%	91%
	02	75%	180%	68%
	03	75%	180%	77%
ETA20Ma	01	17%	18%	18%
	02	22%	20%	24%
	03	19%	21%	23%
ETA20AC	01	16%	17%	19%
	02	15%	17%	17%
	03	17%	22%	19%

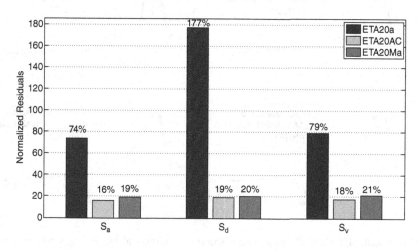

FIGURE 8.7 Normalized residuals of ETA20a, ETA20Ma, and ETA20AC series.

TABLE 8.3
Spectral Matching Simulating Procedure vs. Unconstrained
Nonlinear Optimization

ETEFs Series	Generating Method	Required Analysis Time (seconds)	Average Residuals (%)
ETA20Ma	Spectral matching	130	20
ETA20AC	Nonlinear optimization	18,000	18

Figure 8.7 shows that the directly generated endurance time excitations prove 3% more accurate than the spectral matched ones. In contrast, the generating process by the proposed method is more straightforward and easier. The proposed method took less than 10E-5 portion of the required time for generating new endurance time excitations based on nonlinear unconstrained optimization. Table 8.3 compares the efficiency and the accuracy of spectral matched endurance time excitations and the directly generated excitations. Although this study presents a very fast and accurate simulating approach in comparison with conventional approaches, simulating more accurate endurance time excitations is still of interest in the ET method.

RESULTS

In this section, acceleration spectra of ETA20Ma01 are compared with targets at different times. Figure 8.8 compares ETA20Ma01 and ETA20AC01 acceleration

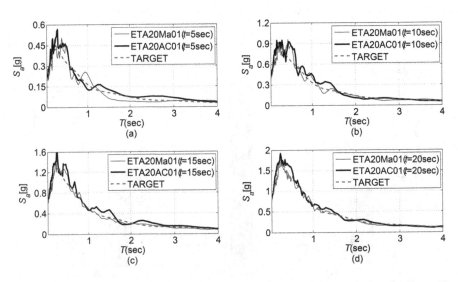

FIGURE 8.8 ETA20Ma01 acceleration spectra vs. targets at (a) $t=5$ seconds (b) $t=10$ seconds, (c) $t=15$ seconds, and (d) $t=20$ seconds.

and displacement spectra with targets at time windows 5, 10, 15, and 20 seconds, respectively. Results show that ETA20Ma01 and ETA20AC01 have somewhat similar accuracy.

REFERENCE

BHRC. (2005). *Iranian* Code *of* Practice *for* Seismic Resistant Design *of* Buildings (3rd Edition). Standard No. 2800-05. Tehran, Iran: Building and Housing Research Center.

9 Generating ETEFs for Direct Response Variability Estimation[1]

REVIEW

Probabilistic seismic analysis has attracted a lot of attention in earthquake engineering in recent years. In this chapter, a method for simulating endurance time excitations which can be applied in probabilistic seismic analysis is presented. Endurance time method as introduced in the previous chapters is suitable for estimation of median seismic response assessment. However, it is also applicable for determining the distribution response parameters which are necessary for probabilistic seismic response assessment. One alternative to achieve this goal is by producing specialized sets of ETEFs as explained in this chapter. Simulated excitations explained in this chapter are also applied in a seismic response assessment to show the practical application of the proposed idea.

PROBLEM FORMULATION

In order to generate probabilistic endurance time excitations, the objective function of Equation (9.1) is defined, which computes discrepancies between endurance time excitations dynamic characteristics and the corresponding ground motions. In this form of objective function, cumulative absolute velocity (CAV) as a duration related parameter has been included in endurance time excitation simulation process. In the equation, F_P denotes the endurance time excitations objective function corresponding to the exceedance probability of P. $S_a(T, t)$, $S_u(T, t)$, and $CAV(t)$ are, respectively, acceleration spectra, displacement spectra, and cumulative absolute velocity (CAV), which are produced by endurance time excitations at time t and period T. In fact, $S_a(T, t)$, $S_u(T, t)$, and $CAV(t)$ are endurance time excitations' dynamic characteristics. $S_a(T, t)$, $S_u(T, t)$, and $CAV(t)$ are computed according to equations presented in Chapter 2. In contrast, $S_{aC,P}$, $S_{uC,P}$, and $CAV_{C,P}$ are ground motions dynamic characteristics which are acceleration spectra, displacement spectra, and CAV targets, respectively.

---

[1] Chapter source: Mashayekhi, M. R., Mirfarhadi, S. A., Estekanchi, H. E., & Vafai, H. (2018). Predicting probabilistic distribution functions of response parameters using the endurance time method. *The Structural Design of Tall and Special Buildings*, *28*(1), e1553.

$$F_P(a_g) = \int_{T_{min}}^{T_{max}} \int_{0}^{t_{max}} \left\{ \begin{array}{l} \left[\dfrac{S_a(T,t) - S_{aC,p}(T,t)}{S_{aC,P}(T,t)} \right]^2 + \\[2ex] \left[\dfrac{S_u(T,t) - S_{uC,P}(T,t)}{S_{uC,P}(T,t)} \right]^2 \\[2ex] \left[\dfrac{CAV(t) - CAV_{C,P}(t)}{CAV_{C,P}(t)} \right]^2 + \\[2ex] \left[S_a(T,t) - S_{aC,P}(T,t) \right]^2 \\[2ex] + \alpha_{S_u} \left[S_u(T,t) - S_{uC,P}(T,t) \right]^2 + \\[2ex] \alpha_{CAV} \left[CAV(t) - CAV_{C,P}(t) \right]^2 \end{array} \right\} dt\, dT \qquad (9.1)$$

where α_{S_u} and α_{CAV} are weight factors which control the contribution of residuals, respectively, associated with displacement spectra and CAV in the objective function. $S_{aC,P}$, $S_{uC,P}$, and $CAV_{C,P}$ are computed according to the following equations:

$$S_{aC,P}(t,T) = g(t) \times S_{a,P}^{\text{target}}(T) \qquad (9.2a)$$

$$S_{uc,P}(t,T) = g(t) \times S_{u,P}^{\text{target}}(T) \qquad (9.2b)$$

$$CAV_{C,P}(t) = g(t) \times CAV_P^{\text{target}} \qquad (9.2c)$$

where $g(t)$ is the *intensifying profile* which is controlling the increasing shape of endurance time excitations acceleration spectra in time. Explanations related to intensifying profile are presented in Chapter 4. $S_{a,P}^{\text{target}}$, $S_{u,P}^{\text{target}}$, and CAV^{target} are normalized ground motion targets acceleration spectra, displacement spectra, and CAV associated with the response exceedance probability of P as computed by the following equations:

$$S_{a,P}^{\text{target}}(T) = \exp\left(\Phi^{-1}(1-P)\sigma_{\ln S_a(T)} + \mu_{\ln S_a(T)} \right) \qquad (9.3a)$$

$$S_{u,P}^{\text{target}}(T) = \exp\left(\Phi^{-1}(1-P)\sigma_{\ln S_u(T)} + \mu_{\ln S_u(T)} \right) \qquad (9.3b)$$

$$CAV_P^{\text{target}} = \exp\left(\Phi^{-1}(1-P)\sigma_{\ln CAV} + \mu_{\ln CAV} \right) \qquad (9.3c)$$

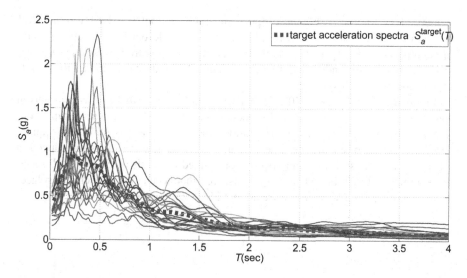

FIGURE 9.1 Acceleration spectra of the recorded ground motions.

where Φ is the cumulative normal distribution, and $S_a(T)$, $S_u(T)$, and CAV are acceleration spectra, displacement spectra, and CAV of normalized ground motions, respectively. μ and σ are the median and the standard deviation operators. It should be noted that peak ground velocity (PGV) is used to normalize the ground motions. The normalization method proposed by FEMA P-695 (FEMA-p695, 2009) is used in this study. It should be noted that far-field ground motions proposed by FEMA P-695 have been used in this study. Acceleration spectra of these ground motions and the associated median along with exceedance probability of 16% and 84% target acceleration spectra are shown in Figure 9.1.

In order to eliminate numerical problems that target response spectra fluctuation may cause in the optimization process, smoothed response spectra are used as targets. Smoothing the response spectra is performed by applying Spline functions.

PROBLEM SOLVING

Discretization is needed in solving the aforementioned objective functions and the type of discretization applied in the solution may directly impact the results. Discretization procedure presented for "lc" series in Chapter 4 is used here. Probabilistic endurance time excitations are simulated in filtered discrete wavelet transform as mentioned in Chapter 3. Trust-Region reflective method is used as optimization algorithm.

SIMULATED EXCITATIONS

In this study, three endurance time excitations for determining the response associated with exceedance probability 16% which are denoted by $ETEF_{16\%}$ are simulated. This endurance time excitation series is named ETA40lc16p in which "40" refers to the ETEF duration, "c" refers to the CAV consistency of simulated ETEFs, and "16" refers to the exceedance probability. This series includes three members, namely {ETA40lc16p01, ETA40lc16p02, and ETA40lc16p03}. The ETA40lc16p01 acceleration time history is shown in Figure 9.2. ETA40lc16p01 acceleration spectra are compared with targets at four periods as shown in Figure 9.3. Acceptable consistency between the endurance time excitations' acceleration spectra and the targets can be observed in this figure.

Figure 9.4 compares ETA40lc16p01 acceleration spectra that are with targets at four times, i.e., $t = 15$, 20, 25, and 30 seconds. This figure shows the acceptable consistency between endurance time excitations spectra and targets. In order to quantify the accuracy of endurance time excitations, normalized residuals introduced in Chapter 4 for "lc" series are employed. Normalized residuals of ETA40lc16P series are summarized in Table 9.1. It can be observed that acceleration spectra, displacement spectra, and CAV accuracy are about 90%, 85%, and 88%, respectively.

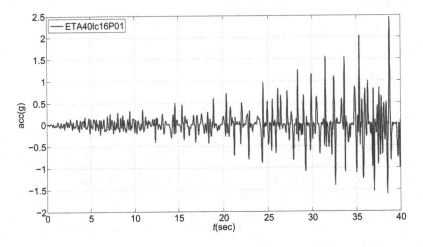

FIGURE 9.2   ETA40lc16P01 acceleration time history.

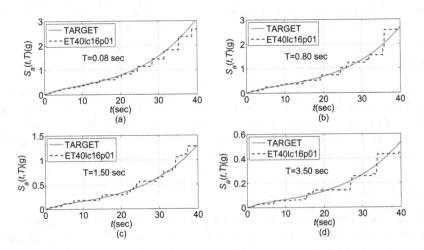

FIGURE 9.3   Comparison of ETA40lc16P01 acceleration spectra and targets at periods: (a) $T = 0.08$ second, (b) $T = 0.80$ second, (c) $T = 1.5$ seconds, and (d) $T = 3.5$ seconds.

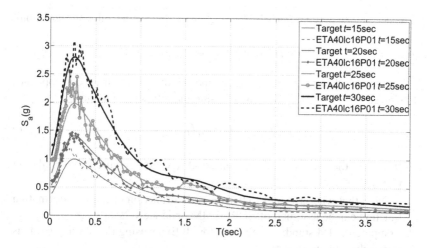

FIGURE 9.4   ETA40lc16P01 acceleration spectra vs. targets at four times: $t = 15, 20, 25$ and 30 seconds.

TABLE 9.1
Normalized Residuals of ETA40lc16P01

		ETEFs Residuals					
		Res_{Sa}		Res_{Su}		Res_{CAV}	
	ETEF						
ETEF Series	**Names**	**Values**	**Average**	**Values**	**Average**	**Values**	**Average**
ETA40lc16P	01	10.5%	10.4%	15.0%	15.2%	12.4%	12.1%
	02	10.4%		14.9%		11.0%	
	03	10.2%		15.8%		13.0%	

SEISMIC RESPONSE PROBABILITY DISTRIBUTION CALCULATIONS

In order to compute seismic structural response distribution parameters compatible with a ground motion suite via the ET method, the following steps must be completed:

1. Normalize ground motions by using PGV.
2. Calculate the natural logarithmic dispersion of ground motion acceleration spectra in the first structural mode period: $\sigma^*_{\ln S_a(T)}$.
3. Scale the $ETEF_P$ by a factor of $\sigma^*_{\ln S_a(T)}/\sigma_{\ln S_a(T)}$, where $\sigma_{\ln S_a(T)}$ is the dispersion used in simulating $ETEF_P s$ and the subscript "P" denotes the exceedance probability of P.
4. Analyze the structure with scaled $ETEF_P s$.

5. Calculate the maximum EDP as a function of time via the following equation):

$$\text{EDP}_P(t) = \Omega(y(t)) = \max(|y(\tau)|) \quad 0 \le \tau \le t \tag{9.4}$$

where $y(t)$ is the EDP response history and Ω is the maximum absolute operator
6. Correlate the ET time t and IM by using the following equation:

$$\text{IM}(t) = g(t) * \exp(\mu_{\ln S_a(T)}) \tag{9.5}$$

$\mu_{\ln S_a(T)}$ is the median acceleration spectra which are used in simulating ETEF$_P$s.

Note that when IM and EDP$_P$ are both a function of t, a transformation function can be assumed of EDP$_P$ vs. IM and vice versa.
7. Compute the IM standard deviation for all EDPs using Equation (9.6). This step is depicted in Figure 9.5.

$$\sigma_{\ln \text{IM}|\text{EDP}} = \frac{\mu_{\ln \text{IM}|\text{EDP}} - \ln(\text{IM}^P(\text{EDP}))}{\Phi^{-1}(1-P)} \tag{9.6}$$

where $\mu_{\ln \text{IM}|\text{EDP}}$ is the median IM calculated by the conventional ET method.
8. Calculate EDP standard deviations which are conditioned to a given IM by the following equation:

$$\sigma_{\ln \text{EDP}|\text{IM}} = \frac{\ln(\text{EDP}^P(\text{IM})) - \mu_{\ln \text{EDP}|\text{IM}}}{\Phi^{-1}(1-P)} \tag{9.7}$$

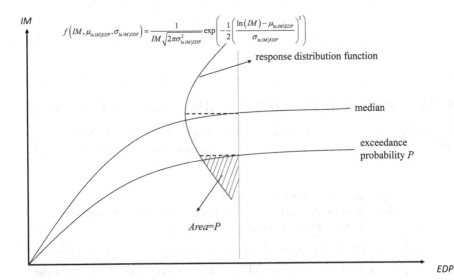

FIGURE 9.5   Calculation of IM standard deviations, conditioned to a given EDP.

where $\mu_{\ln EDP|IM}$ is the mean natural logarithm of the EDPs computed by the conventional ET method.

APPLICATION

So as to show the applicability of the proposed method in this research, EDP probability distributions of three concrete special moment frame structures are computed and then compared with the IDA. It should be mentioned that the IDA was conducted by 200 GMs selected by Heo et al. (2011). In both the IDA and ET methods, the peak interstory drift ratio (PIDR) is considered as the EDP and acceleration spectra in the first structural mode period ($S_a(T)$, where T is the structural first mode period) are considered as the IM. Acceleration spectra of these ground motions are depicted in Figure 9.6. The median acceleration spectrum is also presented.

The proposed methodology is applied to three concrete structures. General characteristics of these three concrete structures are brought in Table 9.2, and configurations are shown in Figure 9.7. These structures are chosen from the Haselton et al. (2008) study.

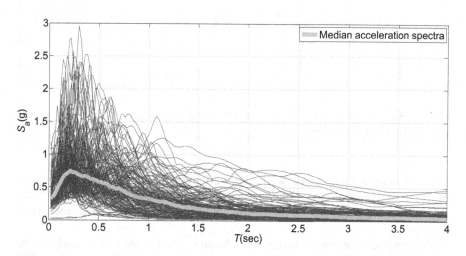

FIGURE 9.6 Acceleration spectra of the GMs used in the IDA analysis.

TABLE 9.2

Properties of Concrete Special Moment Frame Structures (Haselton et al., 2008)

Structure ID (Refer)	No. of Stories	Tributary Area: Gravity/Lateral	Natural Period (s)
1008	4	Space frame (1.0)	0.94
1012	8	Space frame (1.0)	1.80
1014	12	Space frame (1.0)	2.14

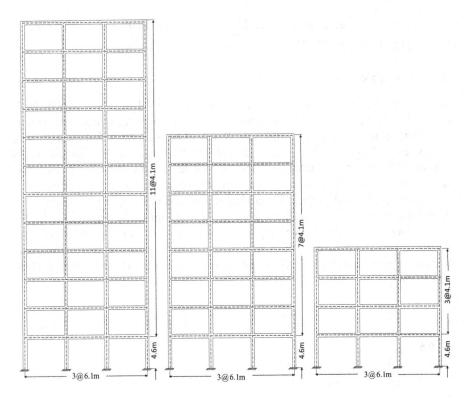

FIGURE 9.7 Geometric specifications of special moment frames (Haselton et al., 2008): (a) 1008, (b) 1012, and (c) 1014.

Since these buildings meet the plan regularity criteria, a single frame is modeled in order to compute the response probability distribution. *Opensees* software (Mazzoni et al., 2006) is used to perform nonlinear dynamic response time history analysis.

Concentrated plasticity which includes beam-column element along with nonlinear rotational springs at both ends is used to model nonlinearity in structures. The schematic modeling procedure is shown in Figure 9.8. Ibarra–Krawinkler model is used to consider flexural stiffness and strength degradation in elements (Ibarra et al., 2005). This behavior model is shown in Figure 9.9. Required parameters in this model are adopted from FEMA-p695 (Federal Emergency Management Agency, 2009). Altoontash model is considered to model panel zone shear deformations (Altoontash, 2004). Readers are referred to Haselton (2006) and Haselton et al. (2011) for more details.

Single IDA curves of model 1008 and corresponding medians along with an exceedance probability of 16% are shown in Figure 9.10.

In the ET method, three ETEFs (ETA40lc01, ETA40lc02, and ETA40lc03) are used to compute the EDP median, and three ETEFs (ETA40lc16p01, ETA40lc16P02, and ETA40lc16p03) are used to compute the 16% exceedance probability EDP.

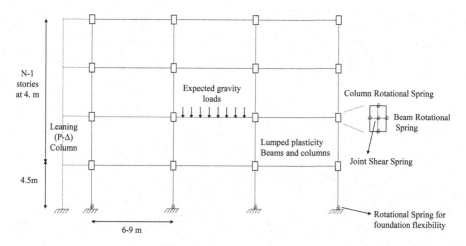

FIGURE 9.8 Schematic model of a nonlinear mathematical moment frame (Haselton, 2006.)

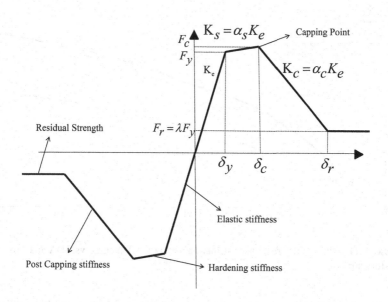

FIGURE 9.9 Backbone curve of the Ibarra–Krawinkler hysteretic model (Ibarra et al., 2005.)

Determining 16% exceedance probability responses is carried out according to the procedure explained in the previous section.

Figures 9.6–9.10 compare ET and IDA method results. These results show that ET method median responses are compatible with median responses of ground motion for all structures (Figures 9.11–9.13).

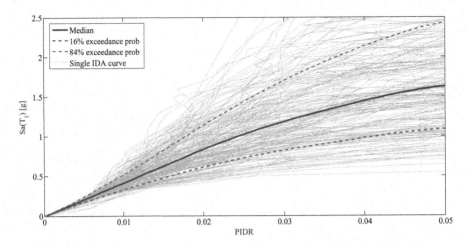

FIGURE 9.10 Single IDA curves of model 1008.

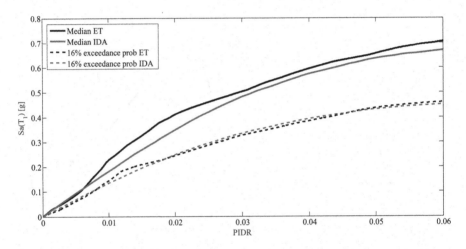

FIGURE 9.11 Probability response estimation using the ET method vs. IDA analysis for the four-story structure.

EDP probabilistic distributions obtained by the ET method for model 1008 at two return periods (475 and 950 years) are presented in Figure 9.14. As it is expected, the probabilistic distribution of responses shifts to right as return periods increase. Moreover, the dispersion of responses also increases as return periods increase, and the proposed method can capture this fact.

Response distribution parameters predicted by the ET method and IDA analysis for the three models at two hazard levels (475 and 950 years) are summarized in Table 9.3. These parameters include the EDP median conditioned to a given IM and the dispersion associated with the natural logarithm of EDPs ($\sigma_{LnEDP|IM}$). The latter parameter is also known as β in the literature. It should be mentioned that the

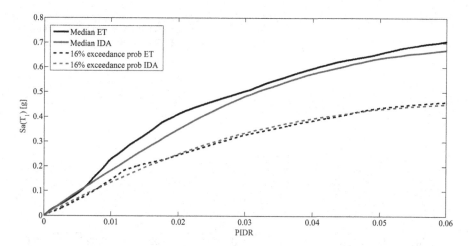

FIGURE 9.12 Probability response estimation using the ET method vs. IDA analysis for the eight-story structure.

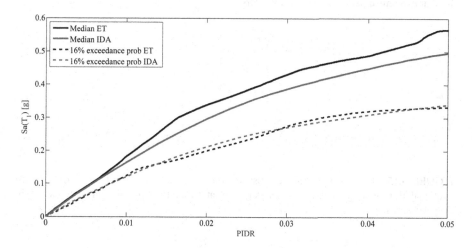

FIGURE 9.13 Probability response estimation using the ET method vs. IDA analysis for the 12-story structure.

differences between the ET method and the IDA analysis results are about 10%–15%. Figure 9.15 compares EDP probabilistic distributions obtained by the ET method and those predicted by conventional time history analysis. It is observed that the ET method predicted well the probabilistic distributions of IDA. In this figure, comparison at the tail part of probability distribution function that is paramount important in the reliability analysis is also provided. This comparison shows that the ET method can also provide a good prediction of the IDA probabilistic response at tail parts.

IM probabilistic distribution functions conditioned to a given EDP are estimated via ET analysis and are then compared with IDA results. In this section, IMs at which

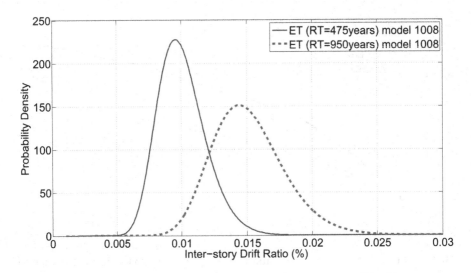

FIGURE 9.14 Response probability distributions obtained by the ET method for model 1008 at two return periods: 475 and 950 years.

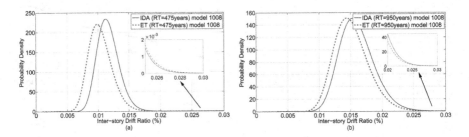

FIGURE 9.15 Response probability distributions obtained by the ET method vs. conventional time history analysis for model 1008 at two return periods: (a) 475 years and (b) 950 years.

TABLE 9.3

Response Distribution Parameters Predicted by the ET Method vs. IDA Analysis

	Response Distribution Parameter							
	475 Years				**950 Years**			
	ET		**IDA**		**ET**		**IDA**	
Models	$\mu_{EDP\|IM}$	$\sigma_{LnEDP\|IM}$	$\mu_{EDP\|IM}$	$\sigma_{LnEDP\|IM}$	$\mu_{EDP\|IM}$	$\sigma_{LnEDP\|IM}$	$\mu_{EDP\|IM}$	$\sigma_{LnEDP\|IM}$
1008	0.0102	0.18	0.0115	0.14	0.0144	0.18	0.0158	0.17
1012	0.0108	0.21	0.0135	0.21	0.0153	0.25	0.0189	0.24
1014	0.0103	0.24	0.0113	0.19	0.0136	0.26	0.0158	0.23

TABLE 9.4

MS Probability Distribution Parameters Conditioned to a Given EDP Obtained by the ET Method vs. IDA

	Distribution Parameters			
	ET		IDA	
Models	μ_{IM}	$\sigma_{Ln(IM)}$	μ_{IM}	$\sigma_{Ln(IM)}$
1008	1.570	0.23	1.630	0.17
1012	0.656	0.16	0.636	0.18
1014	0.566	0.20	0.496	0.19

collapse is occurred are investigated. In this chapter, three criteria are considered for collapse assessment. The first one considers a limit of 0.1 for PIDR, the second one considers the reduction of IDA curve slope (equal or less than 20% of the elastic region slope), and the third one considers the singularity of structural models. Obviously, the minimum value obtained by using these criteria is considered as IM of global collapse of the structure. The results which are summarized in Table 9.4 show an acceptable prediction of the ET method. It should not be ignored that concrete structures have a more complicated behavior in comparison to steel structures (e.g., pinching and degradation in cyclic responses), and hence, such extent of accuracy in prediction by the proposed method is astounding.

REFERENCES

Altoontash, A. (2004). *Simulation and damage models for performance assessment of reinforced concrete beam-column joints.* PhD thesis, Stanford University.

FEMA-p695. (2009). *Quantification of Building Seismic Performance Factors.* Washington, DC: Federal Emergency and Management Agency.

Haselton, C. B. (2006). *Assessing seismic collapse safety of modern reinforced concrete moment frame buildings.* PhD thesis, Stanford University, Stanford, California.

Haselton, C. B., Goulet, C. A., Mitrani-reiser, J., Beck, J. L., Deierlein, G. G., Porter, K. A., ... Ertugrul, T. (2008). *An Assessment to Benchmark the Seismic Performance of a Code-Conforming Reinforced Concrete Moment-Frame Building.* PEER Report 2007/12, Pacific Earthquake Engineering Research Center, University of California, Berkeley, California.

Haselton, C. B., Liel, A. B., Deierlein, G. G., Dean, B. S., & Chou, J. H. (2011). Seismic collapse safety of reinforced concrete buildings. I: Assessment of ductile moment frames. *Journal of Structural Engineering, 137*(4), 481–491. https://doi.org/10.1061/(ASCE)ST.1943-541X.0000318

Heo, Y., Kunnath, S. K., Asce, F., & Abrahamson, N. (2011). Amplitude-scaled versus spectrum-matched ground motions for seismic performance assessment. *Journal of Structural Engineering, 137*(3), 278–288. https://doi.org/10.1061/(ASCE)ST.1943-541X.0000340.

Ibarra, L. F., Medina, R. A., & Krawinkler, H. (2005). Hysteretic models that incorporate strength and stiffness deterioration. *Earthquake Engineering and Structural Dynamics, 34*(12), 1489–1511. https://doi.org/10.1002/eqe.495

Mazzoni, S., Mckenna, F., Scott, M. H., Fenves, G., & Jeremic, B. (2006). *Open System for Earthquake Engineering Simulation.* Berkeley, CA.

10 Optimal Objective Functions for Generating ETEFs[1]

REVIEW

In optimization problems, formulation of the problem is defined in terms of an objective function. In the problem of simulating endurance time excitations, the objective function can be defined in many different ways and forms regarding considered parameters and respective weighting factors. The method for calculating residuals (absolute and/or relative) can also diversify objective functions. This chapter presents a study on the optimal objective function for simulating endurance time excitations. The proposed method for determining optimal objective functions includes quantifying the accuracy and performance of endurance time excitations as a scalar quantity regardless of their generating objective functions. The proposed method can be applied for assessing desired ETEFs and their objective functions of other forms than those studied in this chapter as well.

GENERALIZED OBJECTIVE FUNCTION

The objective function of this optimization problem is defined as in Equation (10.1). It should be noted that this objective function intends to minimize the residuals defined as differences between the endurance time excitations response spectra and targets. This objective function integrates absolute residuals over all times and all periods. The first three terms in the objective function compute residuals in an absolute manner, while the other three terms compute residuals in a relative manner.

---

[1] Chapter source: Mashayekhi, M. R., Estekanchi, H. E., & Vafai, H. (2018). Optimal objective function in simulating endurance time excitations. *Scientia Iranica*, 27(4), 1728–1739.

DOI: 10.1201/9781003216681-10

$$F_{\text{ETEF}}(a_g,\boldsymbol{\alpha}) = \int_0^{T_{\max}} \int_0^{t_{\max}} \left\{ \begin{array}{l} \alpha_a \left[S_a(T,t) - S_{aC}(T,t) \right]^2 \\[1mm] + \alpha_u \left[S_u(T,t) - S_{uC}(T,t) \right]^2 \\[1mm] + \alpha_v \left[S_v(T,t) - S_{vC}(T,t) \right]^2 \\[1mm] + \alpha_{Ra} \left[\dfrac{S_a(T,t) - S_{aC}(T,t)}{S_{aC}(T,t)} \right]^2 \\[3mm] + \alpha_{Ru} \left[\dfrac{S_u(T,t) - S_{uC}(T,t)}{S_{uC}(T,t)} \right]^2 \\[3mm] + \alpha_{Rv} \left[\dfrac{S_v(T,t) - S_{vC}(T,t)}{S_{vC}(T,t)} \right]^2 \end{array} \right\} dt\, dT \qquad (10.1)$$

$t_{\max}$ is the excitation duration and $T_{\max}$ is the maximum period considered in generating. $\boldsymbol{\alpha}$ is the weight vector that is $\boldsymbol{\alpha} = [\alpha_a, \alpha_u, \alpha_v, \alpha_{Ra}, \alpha_{Ru}, \alpha_{Rv}]$. Different weight vectors produce different optimization scenarios. $\boldsymbol{\alpha}$ determines the residuals weight factors of each components in the objective function. For example, when all components of the weight vector are zero except that α_a, it implies that acceleration spectra absolute residuals are considered in the objective function. In this chapter, optimization scenarios are defined according to values of weight vectors. α_a is the factor of acceleration spectra residuals which is computed in absolute manner. This factor is always one. α_{Ra}, α_{Ru}, and α_{Rv} which are factors of acceleration, displacement, and velocity spectra computed in relative manner are either zero or one. This implies whether the relative error of these quantities is considered or not. It should be mentioned because velocity spectra consistency is not considered solely due to the fact that it is less important than displacement spectra. α_u and α_v are the displacement and velocity spectra residuals which are computed in absolute manner. These factors are assigned either zero or one in the literatures. This study introduces a new method for assigning these factors in which the importance of displacement spectra and velocity spectra residuals in the objective function is the same. These factors are computed according to the following equations:

$$\alpha_u = \frac{\displaystyle\int_{T_{\min}}^{T_{\max}} S_a^{\text{TARGET}}(T)\,dT}{\displaystyle\int_{T_{\min}}^{T_{\max}} S_u^{\text{TARGET}}(T)\,dT} \qquad (10.2)$$

$$\alpha_v = \frac{\int_{T_{min}}^{T_{max}} S_a^{TARGET}(T)dT}{\int_{T_{min}}^{T_{max}} S_v^{TARGET}(T)dT} \qquad (10.3)$$

The procedure explained in Chapter 4 is used to simulate endurance time excitations.

FINDING OPTIMAL OBJECTIVE FUNCTION

The proposed method represents a procedure for finding the optimal endurance time excitations simulating objective function. The objective function parameters must be specified at first.

Weight factor vector (α) is considered as endurance time excitations objective function parameters. Different weight factor values define different simulating scenarios. Determination of the best value for this parameter is of concern in generating new endurance time excitations.

In this regard, a criterion must be defined to compare the accuracy of endurance time excitations produced by different scenarios. In this section, the desired criterion is developed. However, the developed criteria might vary for different applications.

The *Normalized Relative Residual* (NRR) for each response spectra quantity is separately computed according to Equations (10.4)–(10.6). NRR integrates residuals at all times and periods. The residuals at each time are integrated at all periods and then are normalized. This normalization method avoids residuals domination where response spectra values are little and division by little numbers occurs.

$$NRR_{S_a} = \frac{1}{t_{max}} \int_0^{t_{max}} \left(\frac{\int_{T_{min}}^{T_{max}} \left| (S_a(T,t) - S_{ac}(T,t)) \right| dT}{\int_{T_{min}}^{T_{max}} S_{ac}(T,t)dT} \right) dt \qquad (10.4)$$

$$NRR_{S_u} = \frac{1}{t_{max}} \int_0^{t_{max}} \left(\frac{\int_{T_{min}}^{T_{max}} \left| (S_u(T,t) - S_{uc}(T,t)) \right| dT}{\int_{T_{min}}^{T_{max}} S_{uc}(T,t)dT} \right) dt \qquad (10.5)$$

$$NRR_{S_v} = \frac{1}{t_{max}} \int_0^{t_{max}} \left(\frac{\int_{T_{min}}^{T_{max}} \left| (S_v(T,t) - S_{vc}(T,t)) \right| dT}{\int_{T_{min}}^{T_{max}} S_{vc}(T,t)dT} \right) dt \qquad (10.6)$$

Total Relative Residual (TRR) is a vector, components of which are normalized relative residuals associated with S_a, S_u, and S_v. This quantity is a vector and thus cannot be used as a comparison tool. Total Relative Cost (TRC) is a comparison criterion which is derived by the inner product of total relative residual and importance vector (I). Importance vector is a unit vector, components of which signify the accuracy importance of each Response Spectra Quantity in simulated endurance time excitations. Schematic computation of TRC is shown in Figure 10.1. TRC is computed according to the following equation:

$$\text{TRC} = \text{TRR.I} = \left[\text{TRR}_{S_a}, \text{TRR}_{Su}, \text{TRR}_{Sv} \right].I \tag{10.7}$$

where "." denotes the inner product operator of two vectors.

The optimal α can be specified by minimizing TRC. The following steps must be taken:

1. Consider a number of optimal possible scenarios; n scenarios numbered SC_i, $i = 1{:}n$, and each scenario has its own α_i. It should be noted that the optimal scenario is selected among these considered scenarios. Therefore, all possible scenarios should be considered. All possible scenarios are discussed in the previous section.
2. Specify an importance vector.
3. Create a number of initial random solutions. m initial random solution numbered X_j, $j = 1{:}m$.
4. Simulate ETEFs(i, j) using the ith scenario and the jth initial random solution. This step is done by optimizing Equation (10.7).
5. Compute total relative cost matrix; each entry of this matrix is the total relative cost associated with a scenario and a random initial motion TRC(i, j), associated with ETEFs(i, j).
6. Normalize total relative cost matrix; entries of each column are divided by the maximum entry of that column. In fact, the total relative cost of different scenarios associated with the same input motion is normalized with respect to each other.
7. Calculate average total relative cost for each scenario by averaging rows of normalized total relative cost matrix. In fact, the total relative cost of each scenario equals the average of normalized total relative cost of that scenario with different random initial motions.
8. Sort different scenarios lowest to highest. The optimal scenario is the one with minimum normalized total relative cost.

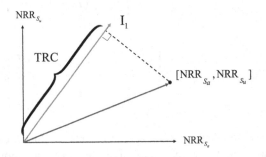

FIGURE 10.1   Computing total relative cost procedure.

APPLICATION

In this section, the method is applied to find the optimum α vector. Nine scenarios are considered; the corresponding α vectors are summarized in Table 10.1. For example, in the first scenario, absolute acceleration spectra residual is considered in the objective function. While in the second scenario, the relative acceleration spectra residual is considered in the objective function. In the third scenario, summation of absolute acceleration and displacement spectra residuals is included in the objective function.

Three initial points are investigated. The first initial point $(X_j\, j = 1)$ is shown in Figure 10.2. It should be mentioned that the units of acceleration, velocity, and displacement are g, g.sec, and g.sec², respectively. The non-zero α_u and α_v values in the objective function are either set at 1 or the value computed as in Equations (10.2) and (10.3). These equations are used as normalization factors.

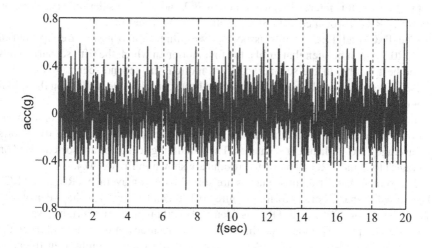

FIGURE 10.2 Initial point of optimization.

TABLE 10.1
Characteristics of Defined Scenarios

Scenarios	α
SC1	[1 0 0 0 0 0]
SC2	[0 0 0 1 0 0]
SC3	[1 1 0 0 0 0]
SC4	[1 13 0 0 0 0]
SC5	[0 0 0 1 1 0]
SC6	[1 1 0 1 1 0]
SC7	[1 1 1 0 0 0]
SC8	[1 13 4 0 0 0]
SC9	[1 13 4 1 1 1]

As shown in Figure 10.2, two importance vectors are considered in order to investigate the sensitivity of the optimal scenario to this quantity, i.e.,

$$I_1 = \begin{bmatrix} \dfrac{1}{\sqrt{3}} & \dfrac{1}{\sqrt{3}} & \dfrac{1}{\sqrt{3}} \end{bmatrix} \text{ and } I_2 = \begin{bmatrix} \dfrac{1}{\sqrt{2}} & \dfrac{1}{\sqrt{2}} & 0 \end{bmatrix}. \text{ TRCs associated with}$$

I_1 and I_2 are denoted as TRC_1 and TRC_2, respectively.

Table 10.2 shows the NRR and TRC associated with these scenarios for the first initial point (X_1).

It can be concluded that the fifth scenario which considers the relative residual of acceleration and the displacement spectra creates better TRC_1 and TRC_2. This scenario creates minimum TRC_1 and TRC_2 in comparison to the other scenarios. TRC_1 of SC5 is 0.239 and is better than other scenarios. It should be noted although SC5 is the best scenario, differences between TRC of SC9, SC2, and SC6 with SC5 are not significant. It is interesting that although SC1 and SC2 consider only acceleration spectra, TRR_{S_a} associated with SC5 is better than SC1 and SC2.

The TRC_1 and TRC_2 results associated with three initial points are reported in Table 10.3. It can be seen that unlike the first motion in which the fifth scenario leads to the best results, in the second and third motion, the second scenario is the best ones. Therefore, it can be concluded that the best scenario is dependent on the initial motion.

In order to find the best scenario regardless of the influence of initial points, average TRC is computed according to step 7 of the proposed method. Figure 10.3 shows the average TRC of different scenarios. The simulating scenarios are sorted based on average TRC_1 and TRC_2. The sorted results are shown in Figure 10.4.

It is shown that different scenarios are sorted based on average TRC_1 and TRC_2. The sorted scenarios are shown in Figure 10.4. It can be seen that the order of scenarios based on TRC_1 and TRC_2 is identical. Although the importance vector is necessary to quantify endurance time excitations accuracy, it does not influence the optimal scenario as well as the order of scenarios from lowest to highest. Justification for this fact is that the information contained in acceleration, displacement, and

TABLE 10.2

TRR and TRC of Optimizations Scenarios

Scenarios	TRR_{S_a}	TRR_{S_d}	TRR_{S_v}	TRC_1	TRC_2
SC1	0.14	0.16	0.18	0.277	0.210
SC2	0.13	0.14	0.15	0.244	0.192
SC3	0.15	0.16	0.19	0.288	0.219
SC4	0.15	0.16	0.17	0.275	0.218
SC5	0.12	0.14	0.15	0.239	0.185
SC6	0.13	0.15	0.15	0.245	0.191
SC7	0.17	0.19	0.21	0.324	0.251
SC8	0.15	0.16	0.17	0.278	0.219
SC9	0.13	0.15	0.14	0.242	0.198

TABLE 10.3

Total Relative Cost for Three Different Initial Motions

Scenarios	$x_1^{(0)}$		$x_2^{(0)}$		$x_3^{(0)}$	
	TRC_1	TRC_2	TRC_1	TRC_2	TRC_1	TRC_2
SC1	0.28	0.21	0.28	0.22	0.34	0.27
SC2	0.24	0.19	0.24	0.18	0.25	0.19
SC3	0.29	0.22	0.28	0.23	0.37	0.29
SC4	0.28	0.22	0.29	0.21	0.27	0.21
SC5	0.24	0.18	0.39	0.32	0.26	0.20
SC6	0.24	0.19	0.30	0.24	0.27	0.21
SC7	0.32	0.25	0.35	0.30	0.36	0.30
SC8	0.28	0.22	0.26	0.22	0.31	0.25
SC9	0.24	0.20	0.25	0.21	0.25	0.21

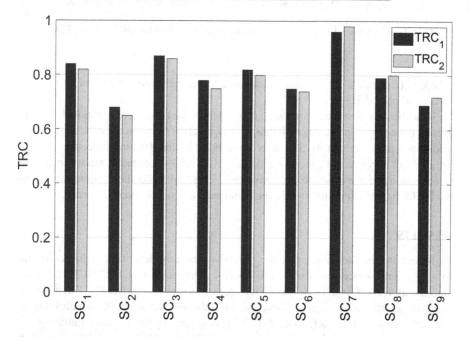

FIGURE 10.3 Total relative cost of different optimization scenarios.

velocity spectra is similar and each can be converted to another by multiplication with angular frequency as shown in Equation (10.8).

Because the endurance time excitations generating process is an optimization-based procedure, the objective function definition influences on the process to find optimum directions toward the best solution.

$$S_u(T) = S_a(T) \times \left(\frac{T}{2\pi}\right)^2 \tag{10.8}$$

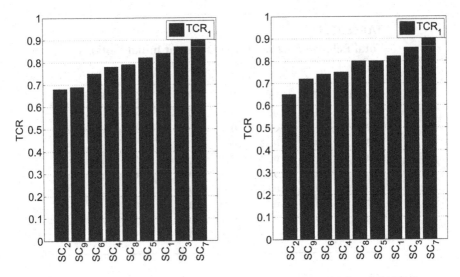

FIGURE 10.4   Sort of different simulating scenarios based on (a) TRC$_1$ and (b) TRC$_2$.

where $S_u(T)$ and $S_a(T)$ are displacement and acceleration spectra of motions at structural period T, respectively.

It can be seen that the second scenario is the optimal objective function definition. Relative acceleration spectra residuals in the objective function create more accurate endurance time excitations. The interesting point is that defining objective function by absolute acceleration spectra residuals has 20% less accuracy than when relative acceleration spectra residuals are considered. This highlights the importance of optimal defining objective function in the endurance time excitations accuracy.

RESULTS

This section aims to investigate the objective function definition influence on the simulated endurance time excitations accuracy. Accuracy of ETA20SC1X2 is compared with ETA20SC2X2. In the ETEFs names, the number after "SC" denotes the scenario number (according to Table 10.1) and the number after "X" denotes the initial motion number. In fact, ETA20S1CX2 is generated by the second simulating scenario, using X$_2$ as initial optimization point. ETA20SC1X2 and ETA20SC2X2 are generated from a similar initial point but different generating scenarios. Therefore, the difference between ETA20SC1X2 and ETA20SC2X2 accuracy comes from those objective functions. First and second scenarios are selected because the first scenario is the current practice for generating endurance time excitations and the second scenario is the optimal simulating scenario determined in the previous section. Targets of these endurance time excitations are based on FEMAP695 [21] far-field record set.

ETA20SC1X2 and ETA20SC2X2 acceleration time histories are shown in Figures 10.5 and 10.6.

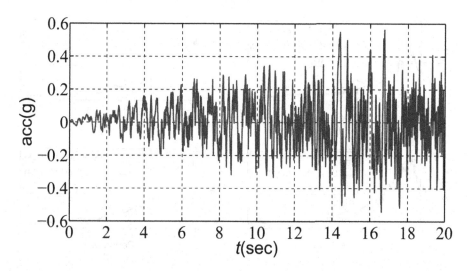

FIGURE 10.5 ETA20SC1X2 acceleration time history.

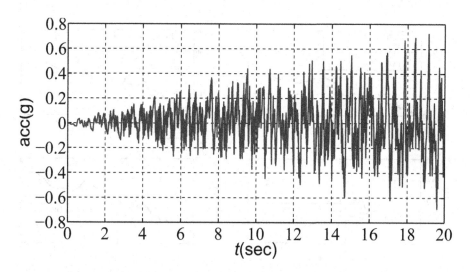

FIGURE 10.6 ETA20SC2X2 acceleration time history.

ETA20SC1X2 and ETA20SC2X2 acceleration spectra are compared with targets at four times, i.e., $t = 5, 10, 15,$ and 20 seconds as shown in Figure 10.7. It can be seen that the discrepancies of ETA20SC1X2 from targets are obviously more than those of ETA20SC2X2.

ETA20SC1X2 and ETA20SC2X2 displacement spectra are compared with targets at four times, i.e., $t = 5, 10, 15,$ and 20 seconds as shown in Figure 10.8. It can be seen that both these ETEFs have acceptable and similar accuracy.

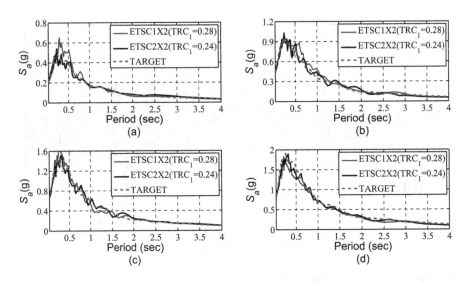

FIGURE 10.7 ETA20SC1X2 and ETA20SC2X2 acceleration spectra vs. targets at times (a) $t=5$ seconds, (b) $t=10$ seconds, (c) $t=15$ seconds, and (d) $t=20$ seconds.

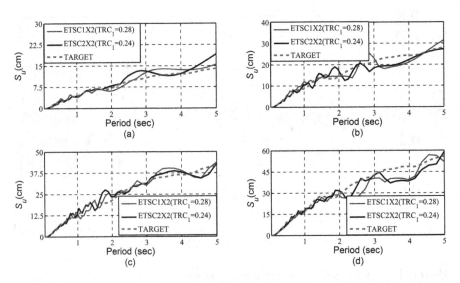

FIGURE 10.8 ETA20SC1X2 and ETA20SC2X2 displacement spectra vs. targets at times (a) $t=5$ seconds, (b) $t=10$ seconds, (c) $t=15$ seconds, and (d) $t=20$ seconds.

ETA20SC1X2 and ETA20SC2X2 velocity spectra are compared with targets at four times, i.e., $t=5$, 10, 15, and 20 seconds as shown in Figure 10.9. It can be seen that both these ETEFs have acceptable and similar accuracy.

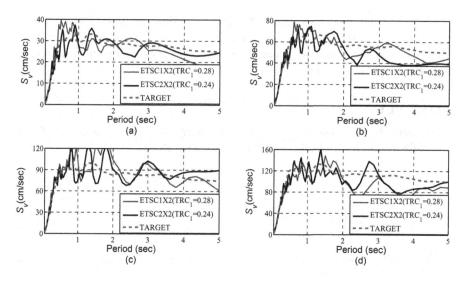

FIGURE 10.9 ETA20SC1X2 and ETA20SC2X2 velocity spectra vs. targets at times (a) $t = 5$ seconds, (b) $t = 10$ seconds, (c) $t = 15$ seconds, and (d) $t = 20$ seconds.

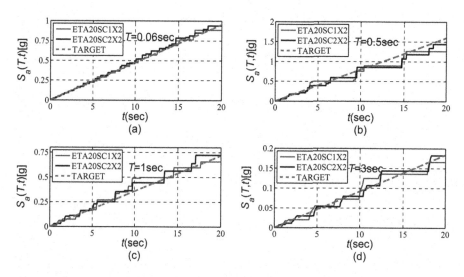

FIGURE 10.10 Comparison ETA20SC1X2 and ETA20SC2X2 acceleration spectra to targets in t at periods (a) $T = 0.06$ second, (b) $T = 0.5$ second, (c) $T = 1$ second, and (d) $T = 3$ seconds.

ETA20SC1X2 and ETA20SC2X2 acceleration spectra are compared with targets with respect to time at four periods: $T = 0.06$, 0.5, 1, and 3 seconds in Figure 10.10. This comparison shows that the ETA20SC2X2 holds higher accuracy.

It can be seen that the objective function definition can appreciably influence the simulated endurance time excitations accuracy. The results show higher accuracy associated with the optimal objective function as compared to the conventional objective function definition.

REFERENCE

FEMAp695. (2009). *Quantification of Building Seismic Performance Factors, FEMA P-695.* Washington, DC: Federal Emergency Management Agency.

Index

Printed in the United States
by Baker & Taylor Publisher Services